BUDDHA SCIENCE

To Annette,
I hope this helps
you to "see" the
elephant.

BUDDHA
SCIENCE

STEVE DAUT

blind elephant books

For permission requests, please address:
Blind Elephant Press
7300 West Joy Road
Dexter, Michigan 48130

Published 2015 by Blind Elephant Press

ISBN 978-1-943290-15-4
Library of Congress Preassigned Control Number: 2016935280

Acknowledgments

I would like to thank everyone who contributed to this effort, beginning with the members of our super-secret philosophical society that I can't mention in public. But thanks Jeff and Bob (and lately, Dick) for some lively discussions. I'd also like to thank those who agreed to take a look at the manuscript and provide input, including Randall Daut, Lila Daut, Brian Hamilton, Sandra Villafuerte, Donna Crudder, Paul Schissler, and especially Peg Tappe, who pointed out the original source for the concept of Flatland.

Thanks to Matt Flickstein for permission to use comments from his course on non-duality and to Robert Wright for permission to use comments from his online course in Buddhism and Modern Psychology. Also, a big thank you to Jeanne Ballew, who provided very helpful editorial comments and suggested a number of changes that significantly strengthened the manuscript and helped me better understand what the book is really about.

Thanks to Carol Blotter for gently steering me back to a path that I had begun many years before.

I also need to especially thank my wife, Susan Lackey, for her vast ability to roll with all of the changes I have gone through since we met. One effect of trying to stay in the moment is that moments continually change, and I tend to change with them. Knowing me, I'm sure Susan won't be surprised that my great and loyal buddy, Chili, gets the last word here. She really has always been a good girl.

Special Thanks

I'd like to thank Ven. Master Haiyun Jimeng and all of the wonderful people at Huayen World for providing a generous grant to support the printing of this book. I encourage you to take a look at their website, visit the monastery, and support their mission to bring Huayen Buddhism to the Western world. Their contact information is as follows:

Website: www.huayenworld-usa.org

Address: Triple Crane Monastery,
          7665 Werkner Road,
          Chelsea, MI 48118, USA

Email: triple.crane@huayenworld.org

CONTENTS

CHAPTER ONE
THE ELEPHANT

Reality is merely an illusion, albeit a very persistent one.
—Albert Einstein

DESPERATELY SEEKING REALITY

What is Reality? Can we ever cut through the craze of the moment, the month or the millennium, bypass biological influences and cultural beliefs and finally arrive at an all-pervasive, never changing core of existence? Is it even possible to understand what a question like that means? Why does it matter? After all, turkeys, trees, and some American presidents have lived out their lives without ever asking these questions. Yet, through the ages, humans have struggled to try and find answers. We need to know what we're here for, and to do that we need to know where "here" is. Are we mere robots, doing the work nature programmed us to do, or do we have the ability to change the reality around us? We seem to be genetically programmed to ask such questions. After all, we are Homo sapiens sapiens, the animal that's not only aware of itself, but is aware that it is aware. Depending on our orientation, we may use religion, philosophy, metaphysics or science—or some combination of these—to try and discover answers.

A scientist looks by viewing these questions as part of a puzzle and trying to understand how the pieces fit together. The philosopher looks inside of his experience to discover logical answers. Sages and religious scholars look to someone or something outside of physical experience to try and make sense of it all.

Why have we looked so hard in so many ways for so many years? Is it because we each feel disconnected from others, from ourselves, from the very flow of life at times? Why do we keep looking when we turn to religion, science, or spirituality, only to find the ultimate answers we seek cannot be found? We seek these answers because we need to feel connected, but the deeper we go into each pursuit, all we find is emptiness, closed minds, or dead ends. We end up feeling more disconnected than ever.

I'm convinced that it's not where we look, but how we are looking. In looking we cling to certainty, to the idea that we can find ultimate answers. Sometimes the different lines of inquiry seem to point in wildly different directions. For instance, we focus on the ways science and religion often clash over their differences, but do they ever agree? In this book, we will attempt to go beyond our either/or mentality to find a deeper level of connection. We'll do it by connecting two of these approaches that may seem at first glance to be wildly different from each other. This book is about the many points where science and the empirical observations of Buddhism seem to agree.

So now you might be thinking, "Wait a minute, Steve! Are you seriously proposing to equate the teachings of a guy who found enlightenment when he got bonked on the head by a coconut, with the findings of a team of theoretical physicists at the Large Hadron Collider, where they bonk subatomic particles into each other at the speed of light?" My response is this: first, it was the Bodhi tree, which has figs, not coconuts; second, I've seen many similarities between these things during my 20-year career as a geophysicist, and during my ongoing exploration of eastern thought. Many other scientists have seen similarities to eastern thought while looking in great detail at specific areas of scientific discovery. And a growing group of mystics are seeing these similarities as well. In fact, we're in the midst of a scientific and cultural revolution that is bringing light to many of the issues I'll be discussing in this book. So, the answer is yes, I see many similarities. As we go along, I believe you will too.

If you toss a dozen great thinkers into a room and ask them what approach we should take to try and understand reality, you'd probably get two dozen answers. But the approach we take to answering the question, and even the language we use to talk about reality,[1] affects the answer so much that we can't even see the insight provided through other approaches. We can be so blinded by our particular point of view that we don't notice the limitations we're imposing. Even when there is a connection, our need to rush to a convenient answer can prevent us from finding it. For instance, western science traditionally limits the concept of reality to those things that can be categorized and measured, while ignoring data that doesn't fit into the conceptual box of materialism.[2] At the same time, mystics and religious fanatics often ignore facts as irrelevant if they believe they have some secret knowledge of the Truth. Is reality purely physical?

Can all things be explained if we can simply find the right tools and ideas to understand them? Or are there deep and mysterious corners of reality that we'll never fully understand? No matter how you go about answering these questions, or what you decide to believe, we all seem to fall into the trap of thinking that if you don't see things the way I do, you must be wrong. You and I must somehow be separate from one another.

We're like the blind men trying to describe an elephant by touching it. This is a story that originated in India, has been told in many different forms, and has found its way into various traditions, including Jain, Buddhist, Sufi, and Hindu.[3] The man who touches the elephant's tail says, "This thing called elephant is like a rope," the man who feels its leg says, "No, it's like a tree," while the man who touches the trunk says, "You're both wrong. It's definitely more like a hose, because it just sprayed me!" Each of these blind men has used his limited perception to develop a fixed concept of what the elephant is like. Since you and I are not one of the blind men in this story and can see the whole picture, we know that each of these men is correct in the observations he's made, but each one is also incorrect because he's limiting his concept of "elephant" to his own personal focus. Each man is too busy clinging to the details of his particular experience to try and find the similarities with what the others are experiencing. If they'd quit arguing, shed their fixed concepts, and try to discover the larger truth that their collective experience is pointing toward, they might be able to piece together a clearer concept of what the elephant actually looks like.

PLANET SILORIAN

Let's imagine a world existing in a golden age, where every type of knowledge has reached an advanced state, but each type has developed in its own individual silo. We will call our world Silorian. The laws and principles that govern Silorian are the same as the "real" one that we currently live in, and the inhabitants have gained complete and detailed knowledge of most of these laws and principles. The inhabitants have achieved this detailed knowledge through a number of separate communities that have passed information along from generation to generation. They have recorded that information perfectly so that nothing was duplicated or forgotten. But what they have not learned is how these laws and principles relate to each other across communities.

There's a community of visual artists who, through the years have developed a deep understanding of beauty and how to evoke it. Another community is built on understanding many generations of religious ideas and

God concepts. They understand the purpose that each of these concepts has served, and how they have evolved over the ages. Yet another community is full of scientists who've built all types of contraptions to make experiments that test the nature of matter and energy, and the working of the universe itself. The community of sages completely understands universal consciousness and non-dual reality. There are communities that know evolution, biological processes, chemistry, and cosmology, each with complete understanding of their own segments of knowledge. The knowledge within each community, although narrow, is so deep and complete that the residents are absolutely satisfied with their lives.

In this world, each community is ringed by a path, and each path is linked to the other communities so that the entire planet is interconnected through this complex of paths. Each community can see its neighbors on the other side of the path, but cannot fathom the activities that are occurring "over there." Although there is no physical reason that the residents couldn't visit their neighbors, they see no particular reason to use the paths, since they have such a satisfying life in each of their communities. Every once in a while, a child will inquire about her neighbor's activities, and she is told that she shouldn't meddle in other people's business, and that she'll understand when she grows up. Why risk exposing herself to the potential danger or ridicule she might find over there?

Most of the children give up after a while, but once in a very great while, a child will wander off down a path into another community. She'll encounter all sorts of things she doesn't understand and she'll meet people who have lived with strange ideas for so long that none is interested or capable of explaining it to her. After a time, this curious one finds her way back to her own community, relieved to be back where people speak the same language she does, and where the world makes sense again.

So, generation after generation, in each of the communities, the sense of us-them grows. It's not that the rest of the world is an enemy to each community, as much as it is simply irrelevant. On this planet, the same truths may be discovered in many different ways, but the various communities never know how similar they may be to their neighbors, or how much more they might discover about their existence by sharing it with others. The paths become a "not what we do" that prevents a shared understanding from ever emerging among the separate communities of Silorian.

The tragedy of Silorian is that the inhabitants have limited their appreciation of the world around them without even knowing they've done it. The artists may appreciate the beauty of a rainbow, but they can't appreciate the beauty of the physical laws that create it. The scientists may be able to describe those laws, but they never even think to ask why the laws came about in the first place. The sages don't attempt to understand how the world actually works because they focus on ethereal matters, believing that the world they can see and touch is inconsequential. They are connected in so many ways, but they never see those connections because they don't look beyond the specific knowledge that they are comfortable with.

A number of years ago, I ran a data processing company that worked mainly with two different types of information: seismic data for oil and gas exploration, and remote sensing imagery from the Landsat satellite program. Since our company worked with both types of data, we also worked with two distinct groups of scientists. I was struck by the fact that each of the two groups was largely unaware of the work that was being done by the other group. They ran in different circles, went to separate conventions and each group generated their own approach to data processing so it could be interpreted in familiar ways. Yet there were many similarities to the two data sets, and it struck me that some cross-fertilization could help open up new potential for each group to gain a deeper understanding of their information. We converted some of the Landsat data to process using seismic algorithms, and our experimentation produced some potentially useful results. But when I tried to explain the process to some of the scientists, I was met with blank stares. Neither group could get out of their particular viewpoint. I didn't pursue it beyond those few small experiments, but I often wonder what would have happened if we could have opened up the boundaries between these two disciplines. And in fact, there are a number of areas with similar data sets—for instance, ground-penetrating radar data for environmental and cultural studies are similar to sonar for subsea mapping, and also to various types of medical imaging. While there are great strides being made in the medical profession, I can't help but wonder how much more could be learned if, for instance, doctors got together with deep sea treasure hunters to talk about how they apply remote imaging techniques. It seems to me that there's a lot to learn by consistent and ongoing work that gleans the best thinking of each profession, and combining this with experimentation to discover new ways to look at and interpret similar types of data.

So planet Silorian is much like the one we inhabit. We divide the world into specialized professions, geographic regions, economic theories, religions and philosophies. At times, these divisions are inadvertent—we simply feel more comfortable with those who share our love of history or art, and so we select our peer groups on this basis. At other times, we do battle to keep the lines of division bright—politics and national interests can take such harsh divisions. But the fact that we divide our knowledge limits us in ways ranging from sad and subtle to enormous and earth-shattering. We simply learn more, experience more, and understand more if we can cross the boundary paths that keep us apart.

THE BOUNDARIES WE BUILD

These boundaries are self-imposed, of course. When we divide our world, we make it more complicated that it needs to be. Rather than seeking the common ground and trying to see how each point of view might fit into the bigger picture, we tend to start from our particular point of view and focus on how it differs from the viewpoint of others. Things could be much simpler if we focused on the similarities. This is not to say that we should ignore the differences, but when we focus too much on what keeps us apart it can be hard to see what brings us together. The Buddha said that if we can recognize our true nature, we can escape the suffering that comes from feeling separate from the world around us. And scientists have recognized that the key to unifying scientific theories is to discover the one very simple underlying principle that drives them.[4] But if this underlying principle doesn't include an explanation of what love is, the emotional impact of the color red, or the events that occurred before the big bang, can it truly be called a universal theory? The story of the blind men is the key that will allow us to begin connecting the ideas of science and those of Buddhism. While we can't physically step outside of our own reality, we can use the story of the blind men to understand that there is a larger level of reality that we cannot fully describe or experience. We'll elaborate on this larger level of reality in the next chapters.

Since my early experiments with the satellite data, the picture has continued to change. As science continues to grapple with the questions of quantum physics, the linkage between science and consciousness cannot be ignored,[5] especially in light of recent discoveries that provide a vital link between subatomic processes and relativity. Prominent personalities such as the Dalai Lama and B. Alan Wallace have said that if long-held spiritual positions are proved false by science, or cannot be replicated in the laboratory, they should be questioned or abandoned.[6,7] In fact, since 1987, the Dalai Lama has been reaching out to the scientific community, resulting in the creation of the Mind and Life Institute, an organization that brings together scientists and contemplatives from around the globe. Many conferences are held that discuss the correlations between science and spirituality, and there are many books and videos that deal with different aspects of this subject. The more you look, the more you realize that we're in a golden age of connection. The information age is morphing into the interconnection age. We're beginning to see linkages that would not have been possible to see even a few years ago.

Although a few prominent thinkers and researchers are becoming more aware of such linkages, we still tend to circle the wagons around concepts that we can grasp and groups that we are at ease with. We want to know where we fit and how it all fits together, so we close out ideas that threaten our comfortable understanding of reality. The things we say and do on a daily basis become our world and our reality, and as time goes on, we become increasingly uncomfortable about stepping out of the mold we create for ourselves. In fact, we're genetically programmed to believe we know the truth, even though we may hold mutually inconsistent information in different parts of our brain.[8] We use our science, or our religion or philosophy, to build a model of the world, and that model becomes the reality that we cling to as the only reality there is. In doing so, we create our own feeling of isolation.

Neither mysticism, nor science, is focused on one single subject. Each is complex, and includes a vast array of details. So we distract ourselves by focusing on the details, and when we do, the big picture fades into the background. We lose the ability to see the assumptions that go into our view of reality, and if these assumptions are challenged, it scares us. If you have ever been through a major earthquake, you know it's terribly disorienting to have the very solid ground beneath your feet suddenly become unstable. You tend to grab onto anything that seems fixed. Finding out that Reality is not what you thought it was is like a psychic earthquake. We try to cling to any little detail that seems real and solid. But sometimes to see more clearly, we need to let go of the details and focus on the larger concepts. So before we delve into the details later in this book, let's first take a broad look at the larger concepts that form the backdrop of both science and Buddhism.

At its heart, science is an approach, a way of investigating observations that applies to many different subjects. Although it's not necessary to believe that all of reality can be explained through the physical world, many scientists work with physical tests and theories on a day-to-day basis, so the tendency for a scientist is to think that reality consists exclusively of things they can touch and measure. In this book, we want to explore the findings of science to see what they can tell us about the nature of reality, but still remain open to the limitations of the scientific method. For our discussions, we will break the subject down into chaos and complexity, physical science, science of life, and science of the mind.

We also have to keep in mind that within each scientific discipline, there are those who focus on the big picture and those who work out the details.

There are mathematicians, theorists, modelers, and experimenters, and each one approaches scientific questions in their own unique way.

Mysticism is a belief that union with the absolute (i.e., God) can be attained through contemplation and self-surrender. This belief can take many forms, but we are focused here only on certain limited aspects of Buddhism. Buddhism itself encompasses various traditions, and throughout the ages as it has grown in geographic extent, it has merged and morphed with other forms, such as Hinduism and Taoism, to produce a complex array of forms and practices. Within each form, there are people who use it as a simple practice for health and wellbeing, people who use it as a philosophy, and people who consider it a religion.

We're not going to get into any of the cosmology, religious, or metaphysical aspects of Buddhism that have been added through the ages. We're more interested in Buddhist observations about the nature of the world and of perceptible reality. This limited approach is actually consistent with the Buddha's admonition to his followers when he said not to be led by the authority of religious texts,[9] and when he emphasized "seeing" - knowing and understanding - rather than religious faith or belief.[10, 11] But we aren't constrained to limiting our view only to the observations of a man who lived 2600 years ago, either. That would be like limiting our discussion of scientific knowledge to the writings of Aristotle.

Frankly, the task is even more difficult because the Buddha left no writings. The first summary of the Buddha's words was written in the Pali language more than four hundred years after his death. Personally, if I want to remember something I heard without writing it down, I can forget it within four minutes, although my wife would claim I probably never listened in the first place. Anyone who has played that parlor game where you pass along a story knows how quickly the story can change. Imagine a game like that lasting 400 years! And to top it off, the Buddha actually spoke Ardhamagadhi.[12] Apparently this crazy sounding language is pretty similar to Pali, but even so, the original writers were translating a verbal tradition from one language into the written form of a different language. What we have today is a translation of a translation. And the translation from Pali to English is no leisurely stroll through the monastery, either. Still, I'm sure the followers and translators did the best they could, so we rely on these texts as the actual teachings of the Buddha.

We also need to understand that trying to form a concept of something as large as science or Buddhism results in a limited view. We can focus on the larger aspects of each topic, but this doesn't capture all of the views that you can have, or the ways it can be used. Individual practitioners utilize the knowledge gained for their own purposes. Although many scientists are also humanitarians concerned about the implications and use of their discoveries, others are willing to subvert their objectivity in order to support a more lavish lifestyle. Ideally, science should not be for sale to the highest bidder. At the same time, the image of all Buddhist monasteries as a place of compassion and mutual support is also idealized in most accounts. In his autobiography, *Journey to Mindfulness*, Bhante Henepola Gunaratana describes brutality, humiliation, and jealousy as part of the Buddhist monastic culture in Sri Lanka.[13]

The point is that science is an amalgam of ideas and concepts built up over the ages, many still under dispute. There is no such thing as "pure" science, and there is also no such thing as "pure" Buddhism. One recent (western) approach based on stripping away Buddhism's metaphysical and religious overtones has dubbed itself "Secular Buddhism."[14] This fits pretty well with the principles I'll be presenting in this book, but I won't use the term again, because the word "secular" comes charged with implications I prefer to avoid. It seems to me the word is one of negation, of division, and the message of this book is connection.

At the same time, it's not possible to discuss all potential connections between every way of looking at the world in a book less than a gazillion pages or so. So, being aware of the limitations and pitfalls of history and language, let's first try to put a box around our topic by building a working definition of religion and metaphysics. Religion and metaphysics exist in the realms of thought and belief. This is tricky when we're trying to understand something about Reality as a whole, because if we're not going to limit reality to the physical, then thoughts and beliefs also exist as part of reality. The purpose of excluding the thought process of metaphysics and the beliefs of religion from our discussion is simply to limit the conversation to a relatively manageable set of topics.

What do we mean by religion? In the broadest sense, we can consider religion to be an attempt to extend the framework of understanding to include an entity or force that is outside the reality that we can observe. One way to describe it is to say that religion envisions an "open" reality that allows for something to exist outside of it, versus a "closed" reality that encompasses everything there is, ever was or will be. In other words, religion is based on the belief that the rules of reality were created by something outside of reality.[15] By the very nature of this belief, there can be no possibility of disproving it within the rules of reality. Also, in this belief, since God is outside of reality, a description of reality in a Newtonian universe does not require God.[16] As long as you can accept the concept of an open reality, there is no inconsistency between a belief in the existence of God and the findings of science. This is because science is purely an attempt to describe a closed reality based on what we can observe and infer from those observations, whereas religion relies on an open reality to explain why they are that way.[17] Science investigates phenomena that can be proven to be false, and there is no scientific way to disprove the existence of a God that exists outside of reality.

To be sure, many people experience religion as a complete inner knowing, a palpable emotional certainty of the presence of God.[18] Empirical knowledge can be defined as knowledge that is gained through sense experience. We will use it to mean observations using human senses that may or may not yet be proven through experimental means. Because the human mind is considered one of our senses, we need to make a distinction between the Buddhist observations of physical phenomenon and the belief that the physical has somehow been created by something outside of the physical realm. How is the inner knowing of religion any different from the empirical observations of Buddhism or science? The distinction is that the observations of Buddhism and science are about reality, and so have the potential to be proven, or disproven, in one way or another. Religious conviction is purely based on faith.[19] In fact, religion must be faith-based, because by our definition, it deals with an entity that exists outside of reality. Reason and logic don't apply here. God may very well exist, by whatever name you wish to use, but such discussions are well beyond the scope of this book. If that sounds like a cop out, well, I can't argue with you. But it makes the book a lot shorter.

Metaphysics, on the other hand, is a search for the rules of reality that can be deduced, but not tested in an experimental context.[20] It is an attempt to explain the fundamental nature of Being and the world that encompasses it.[21] It shares with science the attempt to describe Reality, but the basis for this description is conception rather than observation. As such, it's an intellectual pursuit, one that relies purely on ideas and logic. For instance, say I want to eat an entire bag of chocolate chip cookies, reasoning that my happiness quotient would increase enough to offset the increased risk of a heart attack. Since there is no experimental way to perfectly measure either my heart attack risk or my happiness quotient, no one can prove me wrong.

In principle, metaphysics does not preclude the possibility of an open reality. On the other hand, since metaphysics is based on logic, it allows for the potential that one day, scientific methods and measuring devices may be able to shed light on metaphysical concepts. If they ever find a formula to calculate the net chocolate chip cookie effect, I could be in trouble.

CAN'T WE ALL JUST GET ALONG?

We now have working descriptions of what we aren't going to talk about, so let's try to define what we do want to talk about. Many of the principles that we will be discussing here are not limited to science or Buddhism. Writings on non-duality can be found in many traditions, including Brahmanism, Judaism, Taoism, Christianity, and Native American.[22] Buddhism itself was based on the foundation of Hinduism, and also had some influence on Greek philosophy and perhaps on Gnostic Christianity as well.[23] The Ionians - Pre-Socratic Greek philosophers who lived in the same period as the Buddha - concluded that material reality was all derived from a single essential substance.[24] Scientific philosopher Anaximander of Miletus described a primordial medium that sounds exactly like the modern concept of dark energy.[25] Philosophical discourse regarding the nature of reality continued with the Pythagoreans, Plato, and Aristotle, and the conversations continue today.

As we stand in the middle of the room at this historic cocktail party, we can hear many conversations that inform our understanding of reality and the findings of science. While we are aware of the many other fascinating conversations going on in the room, we can only follow one at a time. So we'll let the others go for the time being and focus our attention on the conversation between Buddhists and scientists, with a few quotations from other traditions that express the principles of Buddhist thought.

A key focus of this book is empirical knowledge, which deals with what we perceive rather than the ideas we form about things. In other words, empirical observations deal with perceptions rather than conceptions. For now, let's just say that perceptions are about observations, and conceptions are about ideas. An example is the observation that I get fatter but not happier every time I eat a whole bag of chocolate chip cookies, versus my continuing idea that it might change the next time I do it. Later on we'll be discussing the distinction between perceptions and conceptions in a bit more detail. So while religion seeks to find meaning beyond our reality, and metaphysics explores conceptions of reality, empirical observations provide perceptions of reality based on direct examination of what our senses tell us.

We will do our best in this book to limit the discussion only to the empirical aspects of Buddhism—those that can actually be verified by experience or science. According to Buddhist principles, this includes direct perception and inference rooted in empirical perception.[26] In order to distinguish between this and the broader religious and metaphysical aspects of Buddhism, we will call this empirical study *Buddha Science*.

BUDDHA SCIENCE

The term *Buddha Science* may be fraught with pitfalls. A theoretical physicist may be tempted at this point to throw this book out the window, believing that the term is oxymoronic or perhaps just moronic in general. At some level, trying to equate the study of navels with solving differential equations seems something like making a pillowcase out of structural steel. I also considered using terms like "Empirical Buddhism" or "Eastern Empiricism," but frankly, *Buddha Science* has a nice ring to it that I hope will translate into more book sales. Of course, calling the book something like "Buddha Meets Schrödinger's' Cat in a Bar" or "Become a Superhero: Get Buddha" would probably sell books too, but I have my limits. Stay tuned for more on the term *Buddha Science* in the next chapter. Oh, and I should also mention in the way of disclaimer that the first initials of *Buddha Science* are an unfortunate coincidence that is totally unintentional, not related to other phrases, pejorative or otherwise, and not subject to speculation as to hidden meanings or opinions.

At any rate, in this book we'll be exploring the parallels between *Buddha Science* and formal science. By formal science, I mean pursuits that follow the scientific method, which will be described in more detail in Chapter 4. By "exploring the parallels" I mean focusing on the similarities. I'm not going to attempt to compare and contrast *Buddha Science* with formal science beyond a short discussion of the differences in basic approach. I'm more interested with where the two pursuits agree rather than where they differ. Along with that approach, we'll be more focused on the big picture than the gritty details of each scientific pursuit. There are a lot of books and websites that provide detail on these parallels, and we'll bring some of those writings into the discussion through the listed references, which are organized by chapter.

I'm taking this approach because it seems to me that we get too focused on the differences in detail. This focus on the details makes it hard for us to see the similarities. The details just get in the way if they cause us to lose track of the bigger picture. The Buddha was describing this tendency when he told a story (described in the Middle-length Discourses) about a man who was shot with a poisoned arrow. The man died because he refused to allow anyone to extract the arrow until he knew all the details about what kind of arrow it was, who shot it, and what kind of bow it came from.

As of this writing, there are approximately seven billion people in the world, which means that there are probably twenty billion different views of reality. This may seem like "fuzzy math," but we'll take a closer look at this later on in the book. And the number of viewpoints is complicated by the fact that, like the blind men, we tend to believe that if our viewpoint is right, the others must be wrong. But like the men and the elephant, if we can focus instead on bringing what we have learned together with the knowledge that others have gained, we all may be able to gain a deeper understanding of a larger reality that none of us can experience as individuals.

When I began thinking about this book, I was in the middle of a year-long workshop with Matt Flickstein that focused on different views of reality. Mr. Flickstein is the author of *The Meditator's Atlas*,[27] and a teacher who provides extended training in eastern thought that includes weekly study lessons and a number of week-long silent meditation retreats. In one of the workshop study lessons, he made the following comments:

> *Truth is one. However, there are two approaches to realizing this truth: via affirma, the positive approach, and via negativa, the negative approach.*
>
> *The positive path is a path of great effort. One has to do whatever is possible to reach the Ultimate Reality. This is a path of seeking, methods, and making progress, all of which takes time. The path is scientific, logical, and proceeds step-by-step.*
>
> *The negative path is concerned with waiting rather than seeking. It is keeping the door to truth open and being receptive. One lets go of any personal intention or goal; it is being surrendered, silent and empty. Time is not needed: realization can occur at any point. This is the path of the mystic. It is mysterious and cannot be explained.*

This is a pretty good summary of the two paths we'll be exploring in this book. The scientific path follows positive action to develop concepts and tests in an effort to pry reality loose from hiding. The path of *Buddha Science*, on the other hand, is based on silent listening, clearing the mind to allow Reality to speak for itself. The main premise of this book is that the approach doesn't matter to reality. Reality is the same, no matter what approach you take to find it, and no matter what aspect of reality your particular path reveals to you.

As I did the research for this book, I began to understand that in recent years, discoveries in quantum physics were driving scientists to take another look at the connection between science and mysticism, beginning with the landmark work first articulated by Dr. Frijof Capra in his 1975 book, *The Tao of Physics*. Since that time, the boundaries between mysticism and science that have dominated western culture for so long have begun to fall, resulting in more integrated approaches to the search for reality. We are in the midst of a scientific revolution, driven not only by this changing view of eastern thought, but also by the revolutionary discoveries of science in the areas of relativity, quantum physics, and chaos/complexity. We are beginning to discover that we are connected in many more ways than we ever imagined.

According to Buddhist teaching, many different fingers can point to the moon, but it's still the same moon. The various paths we take to understand our world are simply fingers pointing to the moon of reality. The goal of the journey is not the fingers—it is the moon. In this book, I will consider two of those fingers: *Buddha Science* and formal science. I believe you will find, as I have, that there are many startling similarities between these two, and both seem to point to the same reality. Let's walk down the path that joins them.

CHAPTER TWO
TWO FINGERS

*I ask you to consider the fact that we live in a web of mystery and have
gotten so used to the fact that we have crossed out the word and replaced
it with one we like better, that one being reality.*
—Stephen King, afterward to *The Colorado Kid*

POINTING AT THE MOON

One of the many variations on Buddhist teaching is that if you point out the
moon to a cat, the cat will look at your finger. I've never tested this observation
with cats, but I know it works with my dog Chili. If I say the word "moon" with
enough enthusiasm while I'm pointing at it, I'm convinced that the concept she
forms in her little doggie mind is "treat." In fact, almost any time I say anything
with enthusiasm, she runs over to the treat jar. Since she's enthusiastic about
treats, apparently the emotion conveys more information than the words. As
humans, we're much more sophisticated about words than Chili is, but we suffer
from the same tendency to form concepts that may or may not have anything to
do with the words, or the intent of the speaker (or author). Since we have many
more words at our disposal than dogs do, we believe we can explain more subtle
and complicated concepts to each other. Add sentence structure and empha-
sis, and there are an almost unlimited number of ways that we can express
ourselves. Perhaps this just gives us an unlimited number of opportunities for
subtle and complicated miscommunication.

Words are hobgoblins that affect the way we see the world around us. We
tend to fixate on the images that words evoke in our minds. As an example,
results that were published online on August 12, 2013 describe experiments

in which researchers at Stanford University used special techniques to blind people before quickly flashing an image of some specific object in front of them. They found that people who were given the name of the object beforehand saw the image more often than those who didn't know what they were looking for.[28] One way to interpret this would be to suggest that when they were given the name beforehand, the test subjects created a mental formation of the object, so when the object appeared they were able to quickly compare it to that mental image. If they didn't have the mental image ready to go, when the object was quickly flashed in front of them there was not enough time to form the image and then use it to recognize the object. Other studies have shown that if you quickly flash images that are unfamiliar, such as playing cards with red spades or black hearts, subjects often completely misidentify what they see.

Anyone who has witnessed an unusual event and compared their experience with other witnesses will recognize this tendency. I remember years ago when I was attending an event in northern Michigan. My friend was driving us along a country road when this animal loped across the road in front of us. The grasses on both side of the road were tall, so we only got a quick glimpse of the animal, but its movements were distinctly lupine. Startled, I asked my friend what it was, and he immediately replied, "Just a deer." I told him what I saw and said I thought it was either a wolf or a coyote, and he looked at me like I was crazy. After arguing back and forth, it was clear that he had decided what he saw and would not listen to reason, and since I'm telling the story, he was wrong. I came to the conclusion that because he was driving, he had less time to focus on the actual critter and therefore simply assigned it to the "deer" image in his mind, an image he was much more used to seeing.

But getting back to the Stanford study, what if they actually saw the object either way? What if they actually perceived an object but they had no mental image with a pre-attached label, so they had no language to describe it? What if they saw something, but they didn't even recognize it as an object? Since our language is object-based, if they could not identify what they saw as an object, perhaps their minds told them it didn't exist. Perhaps what they actually saw was a process instead of an object.

Ok, you might think I'm drinking funny Kool-Aid again, but let's look at it from another direction. The Nootka Indians in the Pacific Northwest of North America use a language that has only one main word category—what we would think of as verbs. But there are no subjects or objects, so everything is described in terms of transient events. The Nootka perceive the world as a process, a stream of events, rather than a collection of objects.[29] Even a book or a mountain is seen as a process, an event that occurs over a particular length of time. It would be interesting to see how the Nootka would respond to an experiment like the one that was conducted at Stanford.

Are objects real, or are they just transient events that mark a convergence of smaller events that we call atoms? Since every language is constrained by a particular viewpoint, is it even possible to describe reality with language? There are those who would answer "no," but I'm trying to write a book here, and I can't draw, so despite its flaws, language is all we have. And although our goal is to understand something about reality, sometimes the best we can do is to describe the fingers, because the moon that we call reality is either a web of mystery, or perhaps a mysterious unending series of events.

In the introduction to their incredible book, *The Quantum and the Lotus*, Matthieu Ricard and Trin Xuan Thuan say that the main difference between the pursuit of knowledge in science versus Buddhism is that in Buddhism, "knowledge is acquired essentially for therapeutic purposes."[30] The Buddha was concerned with the causes of suffering in the world, which he viewed as attachment to our mistaken view of reality. Science, on the other hand, seeks to acquire knowledge for its own sake. Ok, perhaps science acquires knowledge for the greater good of humanity, whatever that means. But the Buddha considered the primary good of humanity to be the relief of suffering through enlightenment, and refused to speculate on the nature of things that did not directly relate to this purpose. He said that arguing about things that are unknowable simply serves to divide us, and enlightenment is about showing us how intimately connected we all are.

So the goals for the accumulation of knowledge are different for science versus *Buddha Science*. Yet in order to understand how our day-to-day perception of the world is in error, we need to explore some other perspectives about reality. As the Buddha taught, everyone must learn the truth for themselves, and the method for gaining this insight is an individual one. Our focus here is about the methods that science and *Buddha Science* use to acquire knowledge about reality, rather than the reasons for seeking this knowledge.

WALKING THE PATHS

Let's begin with science. In general, science begins with inductive reasoning as a method to try and understand how things work. This method is a function of the thinking mind, direct observation, measurement, and trending. Inductive reasoning says that if you make a number of observations that are consistent and repeatable, you can assume that you would probably observe the same thing no matter how many more times you look. For instance, if I let go of an object in midair, it drops to the ground. I can repeat this process with different objects in different places on the earth. Even though I can't test all locations with all objects, according to deductive reasoning, I can logically assume that it will occur with all objects in all places. So we create a rule that says "things fall down." Later on, I'm testing my rule by standing at the edge of a canyon with a strong updraft. I drop an acorn off the cliff, and because of the updraft it actually flies up instead of down, so I have to modify my rule.

In order to understand why the acorn flew up instead of down, I now have to form a concept of why that might have happened. My new rule is that "things fall down except if there is a strong updraft." I try the experiment again from the same cliff with an anvil and of course the anvil falls down. Once again, I have to modify my rule. Then I develop some more experiments based on the new rule to see if it correctly predicts the results of these experiments. In this way, scientists form concepts to explain what we see and experience and then test those concepts. As new information is introduced, the concept is either confirmed for the time being, modified to incorporate the new information, or replaced with a new concept that seems to explain everything that is known at that point in time. Either way, the method is driven by forming concepts around observations.

Although *Buddha Science* also starts with observation, it has a much different view of concepts. It says that concepts get in the way of actually seeing and experiencing the world around us. The methods employed by *Buddha Science* are therefore designed to clear out all of the noise, thoughts, distraction, and details to achieve the silence necessary to experience things as they are. *Buddha Science* says that "doing" gets in the way of "being." It involves eliminating the distraction of thought in order to listen. It says that, although we use concepts to understand something about reality, we can only fully experience reality by getting rid of the concepts that keep us boxed in. While science begins by looking at bits and pieces of the world around us, and then uses inductive reasoning to develop concepts about the larger patterns of reality, *Buddha Science* begins by shedding all concepts and trying to directly experience the larger shape of reality. According to *Buddha Science*, once this shape is experienced, you will also understand how the bits and pieces fit into that shape.

Another way to think about the difference between formal science and *Buddha Science* is to suggest that formal science deals with objective phenomena that can be proven (or disproven) and measured by experimentation, and *Buddha Science* is focused on aspects of reality that can only be felt and understood at the subjective level, but still can be rigorously explored and tested by disciplined mental training.[31] Meditation, for instance, can give practitioners insight into their thoughts and help them reduce the effects of emotions such as anger or resentment.[32] The methods of science and *Buddha Science* can be considered complimentary investigative approaches that can lead to discovering reality.[33] Just as the different blind men each contribute something useful to the concept of elephant, science and *Buddha Science*, taken together, can help us see reality a little more clearly.

On the other hand, *Buddha Science* says that some things are, and will always be, unknowable. Since concepts get in the way of our understanding and we equate "knowing" something with the concept we have of it, we can see or experience things that we will never be able to "know" from an intellectual perspective. Lily Tomlin actually captured the position of *Buddha Science* very well when she said, "Reality is nothing but a collective hunch." Classical science, by contrast, typically takes the position that we can eventually discover all of the principles that govern our reality. But if you think that formal science deals with the known and *Buddha Science* deals with the unknown and intangible, consider the following examples.

Mathematical science is an indispensable tool for most scientific pursuits, and may be thought of as a way to efficiently describe nature.[34] But even this bastion of precision and theoretical purity has its darker side. Mathematicians use the term "irrational" for a number that cannot be precisely calculated— in other words, no one knows, or can ever know, exactly what the number is. When you consider the choice of terms, you can't help but wonder if this isn't science's way of scolding these unruly numbers: "You're being irrational! Settle down and define yourself!" An irrational number is out of the norm, an outlier, not seen in a rational world, right?

Enter the rock star number, pi. This little guy is about 4,000 years old, having first shown up in an Egyptian papyrus scroll around 2,000 BC.[35] He has been featured in a blockbuster novel[36] and a movie.[37] Pi day is celebrated around the world every March 14.[38] As of this writing, pi has been calculated to 12 trillion digits,[39] and still no one knows exactly what the number is.

So contrast this unruly child with its mom, the circle. She has been featured in enigmatic structures from Stonehenge to alien crop circles. She shows up everywhere in nature, from planet profiles to mushroom fairy rings. She is used as the symbol for the great unknown and is celebrated for her perfection. The circle is considered one of the most simple and elegant geometric shapes. Yet circles have their secrets.

In fact, mathematically you can never know everything there is to know about a circle because of her pesky little child, pi. Because of pi, if you know a circle's circumference, or area, or radius, you can't use any of those to exactly calculate the other two. Here we have the most universal shape, the most natural, comfortable and seemingly simple shape we know, and it is mysterious. So is pi irrational, or simply unknowable?

This is only one simple example of a bit of mystery hidden in a seemingly simple and familiar place. Another example is hidden in the very stuff of daily life. Consider the objects around you—the chair you sit on, the walls of your room, the car you drive. These things seem to be real, solid, and distinctly different things. But if you look very closely, every one of them is made up of exactly the same subatomic particles, little bits of matter and energy whirring around with vast empty spaces between them. As the Nootka Indians have observed, the stuff we see as objects are actually events. Golfers say that trees are 90% air, but in fact, it's more like 99.999999999999% empty space,[40] and golf balls are equally vacuous. So why do golf balls bounce off trees with such a resounding thwack? Science explains this in terms of repulsive forces and electron clouds, but there are still many unanswered questions about the nature and origin of these forces.

Science has uncovered many more deep mysteries about reality that we will be exploring in the pages that follow. For now, let's shift gears and take a quick look to see if there is anything tangible in *Buddha Science*.

Let's look at a key practice of *Buddha Science*: mindfulness meditation. Although meditation is not unique to Buddhism, mindfulness meditation is the key technique that has been used by practitioners to clear the mind and allow for the observations of *Buddha Science*. Studies have shown that mindfulness meditation practice can enhance attention and working memory, help students to improve test scores, and can actually effect positive physical changes in the brain.[41] In one example, a rigorous clinical study provided evidence that meditation-based stress reduction also reduced many stress-related symptoms in cancer patients.[42] In a current study, a researcher at New York University is conducting MRI scans of the brains of Buddhist monks to track physical changes in the brain during the practice of deep meditation.[43]

Although not all scientists would agree, formal science and *Buddha Science* generally share the perspective that reality encompasses some elements that are tangible and measurable, and others that are mysterious and cannot be perfectly known or understood. For instance, the study of chaos has shown that natural systems clearly show two distinct types of behavior—those that are predictable and measurable, and those that we cannot fathom or predict. Quantum physics also points to two different worlds—the macro world that we can observe and understand, and the world of the quantum, which has a completely different set of rules.

TWO REALITIES

Buddha Science explains this conundrum with the perception that there are two levels of reality: the dualistic reality of our day-to-day experience, and the ultimate non-dual reality that cannot be fully understood. When we use the term dualistic, we simply mean dividing things into "this" and "that." Some examples are "me" and "not me," "physical" and "non-physical," and "light" and "dark." However, the ultimate level of non-dual reality can't be divided, measured, or compared. Even at the day-to-day level, there's no yardstick to measure any aspect of reality except the yardsticks we create. There was no such thing as a yard until 10th century Saxon King Edgar stretched out his arm and decreed that a yard would be defined as the distance from the tip of his nose to the end of his thumb. Length, time, volume, and anything else we measure, compare, or test, are only known in relation to something else. As you might expect from its name, relativity theory is actually based on the observation that everything is relative.

Since non-dual reality cannot be divided, it cannot be measured, compared, or tested the way we measure and test our day-to-day reality. While we may be able to experience or intuit non-dual reality as a whole, we cannot describe it, and so we also cannot determine the relationship between it and our day-to-day experience. Non-duality cannot be completely expressed in words, just as the exact value of pi cannot be expressed in numbers.

We could say that there are two reasons for this inability to describe non-dual reality. The first is that since reality is larger than any concept, any time we form a concept of what reality might be like, we are excluding some other part of reality. There is no yardstick to measure it against; nothing that reality is "like" to compare it to. So while the blind man could conceive of the elephant's legs as being like a tree, this could only partially help him to understand the concept of elephant. Any conception we have of reality is necessarily only a part of it, a finger pointing toward it, or a map that represents it. To adapt a phrase from a former president Bill Clinton, reality depends on what your definition of "it" is.

The second reason we can't express non-dual reality is related to something we have already discussed—the limitations of our language. Not only does our language divide the world through the subject-verb-object structure,[44] but it also treats words as "elementary units" that can be combined to express anything that needs to be said.[45] As we have seen, this is an idealistic notion, because language itself is built on concepts. Words are associated with concepts, and concepts are imbedded in the structure and forms of language, so the best we can do is to use language to point in the direction of something that is directionless. This may sound like a contradiction. But the contradiction is not an inherent property of reality; it is due to the limitations of the language we have available to describe it.[46] After all, our language allows self-reference that can lead to paradox.[47] Consider the sentence "This sentence is false." The sentence is only true if it is false and only false if it is true.[48] Yet it follows the rules of our language. If language allows us to create such mangled logic within a simple sentence, it is only a crude tool for trying to describe reality itself.

Zen Buddhism uses this property of language to help students think about the nature of reality. One way they do this is through *koans*, which are short paradoxes that cannot be solved with reason. By meditating on this unsolvable riddle, the student is supposed to realize the futility of logic and thereby renounce all thought and achieve enlightenment. Obviously, if this is how Zen Buddhists renounce all thought, they don't have video games. At any rate, probably the most famous of these koans is the simple question, "What is the sound of one hand clapping?" They also attempt to achieve the same result with stories that make you think about the meaning of words. A Zen anecdote that has been passed down over the centuries goes something like this:

> As the Zen master was about to die, the monks gathered around him and asked, "Master, please tell us what life is like so we know what to do when you are gone." Immediately, the master said, "Life is like a river." One of the younger monks asked, "Master, please, what do you mean when you say life is like a river?" After a brief pause, the master replied, "All right. Life is not like a river."

Our Zen master understands that language is only a tool pointing to a concept. If the concept does not lead to understanding, it's not helpful. When the student needed an explanation, it was clear to the master that his analogy wasn't helping the student understand life. It was time to drop the idea.

According to *Buddha Science*, non-dual reality can't be adequately described, but it can be experienced. A dramatic awakening is not necessary in order to experience non-dualism on an ongoing basis.[49] Scientific researchers describe the experience of being inside of the world they are studying, or of experiencing complete understanding of their subject in a way that cannot be fully described in words. A lot of athletes describe being "in the zone" where there is no separation between them and the flow of experience. Though they may try to describe what this experience feels like, the description cannot duplicate the feelings they actually experience. Those who practice meditation on a regular basis often experience heightened connection to their surroundings even when not actively engaged in meditation. That sense of complete connection can be viewed as a form of non-dual experience. It is possible to exist in our day-to-day life and still hold a deep understanding of how intimately connected we are to each other and, in fact, to everything there is.

Although religion is not our focus here, it's worth noting that virtually every religious philosophy in the world refers to an indescribable non-dual reality. Chapter one of the Tao Te Ching describes the experience of non-dual reality (the Tao) as follows:

> *Look, it cannot be seen—it is beyond form.*
> *Listen, it cannot be heard—it is beyond sound.*
> *Grasp, it cannot be held—it is intangible.*
> *These three are indefinable;*
> *Therefore they are joined in one.*[50]

The Indian philosophy of Adviata Vendanta asserts that a positive non-duality centers on a universal self, or Atman. By acquiring knowledge, the practitioner can escape from the bounds of our dualistic world to exist in the non-dual world of pure consciousness.

Even the Bible describes a non-dual reality that existed before the world was separated from God. In the King James Version, John 1:1 says:

In the beginning was the Word, and the Word was with God, and the Word was God.

So in the biblical view, in the beginning there was no separation between God and everything else. Then God created physical reality, which was the first division of the original non-dual reality. This gets back to the idea of an open reality as we discussed in the first chapter. We live in the physical reality, and God, at least as described in the Bible, exists outside of the physical realm. This traditional Christian view may be one reason we have lost our sense of connection. In this view, Christ is the figure who rebuilds this connection and restores the original non-dual relationship with God.

Even though these various views of non-dual reality are very different from each other, they all point to a reality that is larger and different from the one we experience day-to-day. Ultimately, non-dual reality cannot be described as one particular thing or another. In fact, as soon as we try to describe it, the description itself is included within non-dual reality, so our efforts are something like an infinity mirror that keeps repeating the same image within itself, infinitely.

Even in our day-to-day reality, we can see elements of non-duality. For instance, consider the distinction between someone observing an event versus the event that is being observed. Classical science takes the position that we can separate the observer from the thing she is observing.[51] Ok, this seems like common sense as well. After all, I have never lost a tooth watching a hockey game, and I've never gotten a concussion from watching football. But modern physics seems to point to on the conclusion that there is no such thing as an independent observer or observed phenomenon.[52] The distinction between observer and observed could be considered to be purely a human invention.[53] In fact, as we have mentioned, experiments in quantum physics describe two states of existence for quantum systems that are fascinating in their similarity to the two realities of *Buddha Science*. They describe the quantum state, which is a state of unrealized potentiality that can only be described as a pattern of probabilities, in which the classical rules are not followed. The second is the post-measurement state in which the state of the system is shown to meet our concepts of either being a wave or a particle. The very act of observation collapses the system from the quantum state to either a wave or particle. We'll take a closer look at this later on, and we'll also make a few comments about why I don't get black eyes when I watch a boxing match.

So it's not enough to write a book about reality; now we're saying that both science and *Buddha Science* suggest that there are actually two different levels of reality. Great—like one reality isn't confusing enough. Ok, in order to keep it all straight, from now on if I'm talking about the ultimate Reality, the word will always be capitalized, whether I refer to it as non-dual or not. If I'm not specifically talking about this overall, ultimate concept of Reality, the word "reality" will not be capitalized. There is one exception to this rule: If direct quotations from others use the word "reality," it will appear in lower case no matter which reality I think they are referring to. There are some significant challenges to making this distinction between the two levels, but in order to keep the distinction as clear as extremely thin and transparent mud, I'll do my best.

If non-dual Reality is the moon that we seek, but our conceptual framework and language don't allow us to describe this Reality, how do we get at it? We'll do it by trying to describe the conceptual frameworks of science and *Buddha Science* so that we can take a broader perspective, rather than dismissing either of them based on our particular viewpoint. Albert Einstein illustrated the idea of working within a conceptual framework when he described a fictional place we will call Flatworld.[54] Flatworld is a spherical world inhabited by beings that live on the surface of this world. Although we, as outside observers, can see that the world is spherical from our three-dimensional perspective, the inhabitants can only experience two dimensions. So they can recognize lines and circles and make calculations that relate to two-dimensional objects, but since they can't recognize three dimensions, they have no way to directly experience the sphere that they live on. Still, being curious beings, they employ their scientists to measure and map their universe. The scientists put a stake in the ground, and start measuring. The first thing they find is that if they start out in any direction and keep going in a straight line, they always end up back at the stake, and the distance they measure is exactly the same. Next, in a flash of insight, one of the scientists says, "I wonder what would happen if we walked in a circle?" So, starting at the stake, they walk in small circles, then larger and larger ones, always starting and ending at the stake. They find that the largest circle possible is just a little bit shorter than the path that goes straight. As we can see from our perspective, the small circles they measure are equivalent to lines of latitude. The smallest circles are like measuring lines of latitude close to the north or south pole, and the as the circles grow larger it is like measuring latitude closer to the equator. The straight path measures the actual circumference of their universe. The scientists all decide this is great fun, and a healthy way to spend taxpayers' money, but it really isn't teaching them as much as they need to know.

These scientists are pretty familiar with circles, so they decide that since they have measured the circumference of various circles, they might as well measure the diameters as well. After all, even though this is a fantasy world, they don't know the exact value of pi either, so they can't perfectly calculate what the diameters will be. So they start with the smallest circle, paint a green line along this path, and put in another stake exactly halfway along this path. Then they go back to the stake, turn 90 degrees, and paint a red line straight between the two stakes. They measure the length of this red line to get the diameter. They find that the ratio between the circumference and the diameter is pretty close to pi. But as they repeat this process for larger and larger circles, they find a very curious thing. They find that as the circumference of the circles increase, the ratio between the circumference and diameter decreases, because instead of measuring the diameter along a straight line they are measuring it along the curved surface of the sphere.[55] By the time they measure along the straight path (the actual circumference of the universe) they find the circumference is exactly twice the diameter. This would be like measuring the distance around earth's equator and comparing it with a line that starts at the equator, goes through the North Pole, and ends at the equator on the other side of the earth. If the earth were a perfect sphere the path through the pole would be exactly half of the distance around the equator.

By repeating these experiments a number of times in different places, the Flatworld scientists would be able to deduce that they live on a "three-dimensional circle." So even though they might not be able to envision what a sphere looks like, they would be able to determine through careful analysis of the ratios they find, that their universe is a circle in three dimensions—in other words, a sphere. At that point, the Flatworld politicians would fight over whether circles were just a ploy to get more research money from the unsuspecting populace, and they'd pass laws giving equal status to squares.

Ok, enough of that. Let's get back to our world, where politicians don't fight over such trivial matters. We don't really know how many dimensions our Reality may occupy, but we know from relativity theory there are at least four—the three dimensions of space plus the dimension of time. How would Reality look if we were able to step outside of it and see its shape? Just like the inhabitants of Flatworld, we exist inside of our Reality so we are limited by our frame of reference. But that doesn't mean we can't get some sense of what that Reality might look like. Given the limitations we have described, let us try to begin the process by exploring non-dual Reality in a general way.

As we have said, in the broadest sense, non-duality can be understood as indivisible Reality. In other words, there is only the whole of Reality, though Reality is experienced in different ways by each of us,[56] much like the blind men and the elephant. So what can we observe that leads us to the notion of non-duality? First, let's look at the way we experience the world. Through our senses, we perceive something. This perception exists within Reality. Then our mind takes over and we decide this is something that exists independently in the world, and we decide to call it a cup. When we form the concept of "cup," we also form the concept of "not cup," so conceptually we have just divided Reality. But does the cup actually exist without the rest of Reality?

Reality hasn't changed when we formed the concept of "cup." The only thing that happened is that we formed a concept. When we formed that concept, we conceptually divided Reality into "cup" and "not cup"—we didn't divide Reality itself, but only the way we were thinking about it. Conceptually, nothing exists except in relation to its opposite. There is no hot without cold, up without down, happy without sad. If everything in the world were one temperature, the concept of hot or cold, or even the concept of temperature itself, would be meaningless. We experience different temperatures, of course, but how do we determine what is hot and what is cold? If you live in Key West and the temperature drops to 50 degrees Fahrenheit, you would experience that as cold. But if you lived in

Michigan in the winter and the temperature rises to the same 50 degrees, you would find it to be very warm. So the concept of warm or cold is only meaningful in relation to something else. In fact, even the concept of 50 degrees only exists in relation to a thermometer. But Reality itself is not limited by the concepts we have about it. Hot and cold cannot exist as independent properties, but only in a relative sense, in relation to each other.

Experiencing the world as a collection of unrelated items and concepts is dualistic. In fact, concepts are merely illusions that divide non-dual Reality. Any time you form a concept, you also form its negation. We divide Reality conceptually, usually without even knowing that we are doing it. But such divisions do not actually exist, except in our minds. Our sense of separation is an illusion. In the day-to-day world, we appear to see a divided Reality because of the concepts we use to try and understand it. For instance, when we see a book, we conceive of it as an object that exists independent of all other objects in the universe. But how could a book exist without the tree that was used to make the pages, or the sun and soil that made the tree grow, or the author of the book? This observation will be developed more fully in the next chapter, but for the moment we can say that if you explore this observation fully, you will find that nothing exists without everything else. We are all are integral to the universe, imbedded in the flow. Reality cannot be divided, except in our minds. And although we don't fully understand the relationship between matter and mind, seeing these two as separate things is an illusion. According to the Buddhist principle of "dependent origination," even matter and mind are co-dependent.[57] Reality is non-dual.

Ironically, one way to try and understand Reality is to form concepts about it, as long as we don't make the mistake of thinking that the concepts we form are Reality itself. One concept from science that is useful for trying to understand indivisible Reality is the concept of infinity, which we will say is the quality of having no limits. Infinity is an integral part of modern mathematics, particularly of calculus. Calculus itself is the basis for many of the scientific models that we use to interpret the universe and predict events. Because of this, science could not exist as it is today without the concept of infinity.

Infinity is simply what it is. Half of infinity is still infinity. Twenty times infinity is simply infinity. Like non-dual Reality, infinity cannot be divided. If all of a sudden, half of Reality disappeared into nothingness, what would be left would be Reality. If we put on a different pair of glasses and look into the infinity mirror, we have changed the image in the mirror, but the mirror itself has not changed.

Non-dual Reality is the backdrop, the basis from which many of the principles of *Buddha Science* spring up. We view our world through a lens of duality, and we come to believe in separate beings and phenomena occupying an objective universe. But if non-dual Reality is indivisible, these divisions are artificial no matter how real they seem to us from our day-to-day point of view. Separate, independent beings and phenomena cannot exist. The disconnect between these two levels of reality is the basis for our existential suffering, and *Buddha Science* has developed certain principles to help us understand both levels of reality and the relationship between them. So now let's take a closer look at these principles.

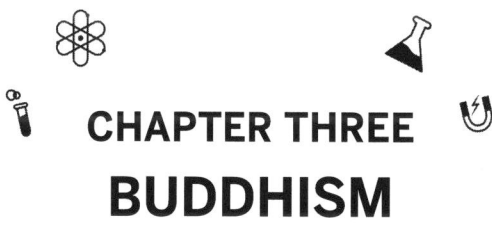

CHAPTER THREE
BUDDHISM

*A concept is anything with a skin around it, some sort of
boundary dividing something from something else.*
—Steve Hagan, in *Buddhism Plain & Simple*

BUDDHA SCIENCE

One finger pointing to Truth is Buddhism. Frankly, I struggled for some time about even using the term Buddhism, because it has become fraught with so much baggage that the term itself is nearly meaningless. Or perhaps a better way to say it is that the term is, for most people, extremely meaning-full. Buddhism has a long and varied history since its founding some 2,600 years ago. For some, it has become synonymous with meditative practice, for others a retreat from the ills of life, and for some it embodies gods and goddesses such as Medicine Buddha, who presides over earthly healing, and Tara, goddess of success.[58] Much like some religions, it has split and evolved, in some cases become morphed and mangled, and has come to mean many different things to many different people around the globe. The term Buddhism actually embodies its own teachings—it has no unchanging essence.

Because the term is subject to such broad interpretation, those who discuss it generally find it necessary to define it, or use another term. In his excellent book, *Buddhism without Beliefs*, Stephen Bachelor used the term "dharma practice" because this term suggests a course of action rather than a belief system, stating that "The four ennobling truths are not propositions to believe; they are challenges to act."[59] The four noble truths are the key component of Buddhism, as they define the basic problem of human suffering and prescribe a pathway out of this problem. They are based on observations about Reality that we will be discussing in this chapter. The four noble truths can be stated as follows:

- First noble truth—Life is unsatisfactory, and causes suffering and anxiety

- Second noble truth—We suffer because all things are impermanent, yet we crave pleasant things and push away unpleasant things

- Third noble truth—We can escape the suffering

- Fourth noble truth—Outlines a path that leads to escape from suffering

Siddhārtha Gautama, the man who is known as the original Buddha, spoke of his enlightenment. But by this, he simply meant that he came to thoroughly understand "the human problem, its origin, its ramifications, and its solution."[60] The four noble truths are a summation of this understanding and how to achieve it. Enlightenment is simply "seeing things as they are rather than as we wish or believe them to be."[61] Gautama repeatedly admonished his followers not to be satisfied with scriptures, teachings, or pronouncements of others, but to weigh for themselves what they determine to be true. In the first chapter, we described empirical observation as a technique to determine what is true, which is something that *Buddha Science* shares with formal science. In fact, metaphysician Jeffrey Grupp has said that "Buddhism is an empirical study … and therefore, Buddhism is a scientific pursuit."[62]

This empirical aspect of Buddhism is best summed up in these words that the Good Reads website has attributed to the Buddha:

Do not believe in anything simply because you have heard it. Do not believe in anything simply because it is spoken and rumored by many. Do not believe in anything simply because it is found written in your religious books. Do not believe in anything merely on the authority of your teachers and elders. Do not believe in traditions because they have been handed down for many generations. But after observation and analysis, when you find that anything agrees with reason and is conducive to the good and benefit of one and all, then accept it and live up to it. [63]

In this book, we are looking at the empirical observations made by Gautama and other practitioners. How is it possible to sort out belief from observation when the observations are made by sitting on a pillow and chanting? In the first place, as we will see, meditation is not the only method that *Buddha Science* uses. In the second place, the whole idea of the book is to see where science and Buddhism agree, so as the Dalai Lama says, if these observations don't hold up to scientific scrutiny they won't make the cut. Whether you look at it as science confirming the observations of Buddhism, or Buddhism echoing the findings of formal science, Buddhists have made many incredibly accurate observations that apply even down to the molecular level and out to the scale of the universe itself, simply through deep meditative practice and observations of the nature of mind, and of the world around them.

Buddhist tradition describes three types of "facts" and the method used to qualify them as factual:[64]

- Direct perception, known through the five senses and mental perception

- Hidden facts, known by reliable inference

- Extremely hidden facts, known by reliable testimony

So these "facts" are what we will be referring to as empirical observations. In other words, I am interested only in the insights from Buddhism that meet the Buddhist concept of "factual." Yet, the term Buddhism can never be reduced to a few simple principles when it encompasses such a diverse and richly historical tradition, because for many people, it is a religion, a philosophy, a practice, a call to action, or many other things. So true to the approach suggested by Stephen Bachelor (*Buddhism Without Beliefs*), I will not be calling these the principles of Buddhism, but the observations of *Buddha Science*.

First we need to develop a basic understanding of how, in Buddhist teaching, accurate observations arise. By a process that varies across different schools and traditions, the first work of a Buddhist is to see things as they are, without the distorting influence of concepts - thoughts, expectations, or ideas.[65] In this process, the practitioner moves from being "asleep" to the Reality of existence and awakens into awareness of everything that is occurring in each moment. When the practitioner has achieved a state in which this awareness is ongoing, they are said to be enlightened. Note that this is not some magical process, in which the Fairy GodBuddha descends from the firmament and conks you on the head with a limb from the Bodhi tree. For most, it's a gradual process, an incremental series of changes in perception and of the way they relate to the world. Yet many seem to believe enlightenment is always some sort of magic, an "aha" moment where we suddenly understand everything. This happens to some fortunate few, but it is the exception, not the rule. Again recalling the power of words, perhaps the discovery of electricity and the ability to turn on a light instantaneously has led us to expect enlightenment to occur just as quickly and magically.

It is not that people who have achieved some level of enlightenment never form concepts, but they are fully aware that these mental formations are not Reality. The mental impression of Reality is never Reality itself. According to Buddhism, this awareness can only be achieved by experiencing it personally. By training the mind to avoid the distorting influence of mental formations, A Buddha Scientist can make accurate empirical observations.

The key process used to develop this undistorted view is meditation. Two basic forms of Buddhist meditation are concentration and insight meditation. Concentration meditation helps the practitioner to achieve mental focus through concentration on a particular phenomenon, such as the breath. By simply observing the breath, by focusing on the sensation of air passing in and out through the nostrils or the rise and fall of the chest, the practitioner can train themselves to pay attention to the moment, letting all concepts fall away. Once a certain level of concentration is achieved, the practitioner practices insight meditation by opening their focus to observe everything that is happening in the moment without attaching concepts or expectations. Consistent application of these seemingly simple practices can result in a completely new way of observing Reality.

These observations are not made from the same perspective as those that might be made by a scientist, because a Buddhist makes their observations from the "inside," instead of from the "outside." By this, I mean that the Buddhist perspective is that we exist within Reality, so any observation we can possibly make is from the "inside" and is therefore subjective. Traditional science, on the other hand, says that we can observe aspects of Reality from the "outside," so we can make unbiased, objective observations.

From the perspective of *Buddha Science*, the most effective way to discover Reality is to fully embrace and inhabit our inside viewpoint. This is, in fact, a theme that we see repeated in many ways every day. Athletes describe "being in the zone" when they are performing at their best, and the very concept of empathy involves inhabiting the world through the perspective of another. Our favorite fictional detectives often solve the crime by getting inside the mind and emotions of the criminal. The actors we admire most describe the extreme methods they use to inhabit a character. So from the perspective of *Buddha Science*, we live life from the inside, and the challenge is to avoid getting so enmeshed in this perspective that we cling to it as if it's Reality itself. This ability to make accurate observations while inhabiting the subjective view is addressed in *Buddha Science* through rigorous mental training.[66] Meditation, as we described above, is a key component of that mental training.

In the next chapter, we will add more detail to the scientific perspective that empirical observations can be made from the "outside," and can therefore be objective. There is actually an interesting corollary between the outside, objective viewpoint of science and the inside, subjective viewpoint of *Buddha Science*. While traditional science says that you can separate the observer from the observed and therefore be objective in your observations, the Buddhist principle of non-duality says there is no separation between the observer and the observed. On the other hand, Buddhism also involves the principle that it is an illusion to think that there is a permanent, unchanging self. If there is no self to observe, but observations are simply a process that occurs, then observations in themselves are simply objective phenomena. So from either the scientific or Buddhistic perspective, something objective is occurring. Ok, I'll admit this is kind of a convoluted point, but this distinction in perspective is one of the key differences between the pointing fingers of Buddhism and science. Neither perspective provides the whole story. Perspective is not Reality.

So getting back to the observations of *Buddha Science*, how do we distinguish between observations and simply personal conclusions? To try to understand the distinction, we need to understand the difference between perception and conception. One way to describe the distinction goes something like this: our senses record information, which is converted into perceptions by our mind, and then the intellect converts these perceptions into concepts or ideas.[67] *Buddha Science* says that perceptions are necessary for accurate observation, and concepts get in the way of it. Although concepts can be useful for one level of understanding, they can distort accurate observation, especially if we make the mistake of thinking that our concepts truly reflect non-dual Reality. We have already discussed the two levels of reality, so let's start by seeing how, according to *Buddha Science*, the two levels relate to each other.

In Chapter 5, we'll be discussing the concept of strange attractors. Attractors, in the language of chaos science, can be thought of as a shape that contains all of the possible states that a physical system can take. The shape, or envelope, never changes. But within this shape, the system undergoes constant change from one state to another. A good example is the weather. The weather is constantly changing, but we know it will be contained within the envelope of the atmosphere. Within this envelope, weather will form certain patterns. There will be vortexes and fronts, wet hot air will rise to form clouds, cold air will sink to form downdrafts, and these patterns will continue to occur, even though they will never exactly repeat themselves. This is also one way to think about the two levels of reality—ultimate Reality as the attractor, which is the broad shape of Reality that is timeless and unchanging, and the smaller patterns of reality that we see at the day-to-day level, which constantly change and flow. We could say that non-dual Reality is the container that holds all of the individual realities that each of us experiences, just as earth's atmosphere is the container that holds all of the weather patterns we experience.

WHAT BUDDHA SCIENCE SEES

The broad shape of Reality includes the dimension of time, so the shape itself is timeless. At the day-to-day level, we experience time as a linear element that changes and flows, but at the ultimate, non-dual level, time has no meaning. Using the analogy of the men and the elephant, we are blind to the wholeness of time. We may be able to sense the shape of the whole, but we can never completely understand it by seeing it from the "inside." Like the beings on Flatland, we have a restricted view of Reality, so we piece together the larger picture by carefully observing what we see from our point of view inside of day-to-day reality. The main observations of *Buddha Science* flow from observations made from the inside. The observations we will be discussing are interconnection, karma, impermanence, and illusion.

The observation of interconnection says simply that everything is connected to everything else. Although we seem to experience the world around us as a collection of unrelated items, if we look more closely and see what is actually there, we will realize that everything is connected. Thich Nhat Hanh uses the example of a sheet of paper, explaining that without clouds there would be no water, and without water there would be no trees to make the paper.[68] An example that I find particularly compelling is that of wine. Although a "pure" Buddhist might berate me for using an alcoholic beverage to discuss Buddhism, I am using the example in a purely metaphorical way. And I do know a little something about wine. Drinking it, anyway.

After all, what makes up a bottle of wine? Is it the bottle itself? Certainly, the wine would not be what it is without a container to keep the air out, and the color of the bottle has a major effect on how well the wine will age. What about the cork? A bad cork will not allow a wine to age without going sour. So the quality of the wine that you open today depends on the bottle and the cork, as well as the way it was stored.

It is true that the wine is the grapes? Certainly it would not be the wine you taste without the particular type of grapes that were grown, but even with grapes that are of the same variety, the wine that is produced depends on the temperature, the humidity, the amount of sun they received, the soil, the time they were harvested, and the geographic location where the vines were grown. Wine makers and wine tasters alike know that the quality of wine varies greatly from one growing season to the next. In addition, grapes of one variety are often grafted onto root stock from another variety more suited to the soils and climate of a region, so another effect is the ability of the root stock to deliver particular nutrients to the vine. And of course, we can't forget the skill of the vintner, the type of barrels in which the wine was aged, the length of time it was allowed to age, filtration techniques, and numerous other factors around the conversion of grapes into wine.

The skill of the vintner, of course, also depends on training and experience. What wines have they tasted that allowed them to make all of the particular choices they have made in producing this particular bottle of wine? What instruction have they received, and what advice have they been given?

Would the wine exist without the vintner, or the cork of the bottle, or the grapes, or the soil, or the sun? Would the soil exist without erosion breaking down rocks, or leaves falling from trees to create humus, or worms to mix the soil, or rain to add moisture? The soil is a product of processes that are dependent on the dynamic systems of the earth and the sun as they exist today. How did the earth and the sun become what they are today?

Actually once it is bottled, bought, and makes its way into our glass, we experience the wine as a thing by itself, unique for all the factors of one particular bottle of wine, for the nose and the color, the taste and texture. But the wine is only exactly the wine we experience because of all the factors that lead to the moment we touch it to our lips. All of the factors described so far are not the end of the story. Our experience of that wine will be different tomorrow, depending on the food we pair it with or the state of our taste buds. Anyone who has tasted wine can tell you that certain wines taste dramatically different if you let them breathe, decant them, or pour them through an aerator. Even the place on our tongue that we are using to taste the wine will affect the flavors we experience!

The wine only exists as what it is because of everything else that exists in the particular moment that we drink it, including the factors that make the drinker the person they are at that moment. Even that momentary person is dependent on the causes and conditions that led to that particular moment. And the first sip tastes different from the next because of the effect the wine has on our taste buds. We experience wine as a constantly changing process, rather than a fixed and permanent "thing." Each moment of tasting connects us to every moment that went before.

Breathing is also something that we experience as a constantly changing process. Enrico Fermi, the Italian physicist who worked on the first nuclear reactor and is one of the men referred to as a "father of the atomic bomb," posed various thought experiments for his university students. One such though experiment was, "When you take a single breath, how many molecules of gas you intake would have come from the dying breath of Caesar?"[69] By making a number of reasonable assumptions it has been demonstrated many times that the answer is, "At least one." One basic assumption is that 2,000 years is sufficient time for gas molecules to be evenly distributed through the atmosphere, which means that you are also breathing in the gas molecules from anyone who lived then or before. With the breath you take right now, you are sharing the use of molecules with not only Julius Caesar, but also Mark Anthony, Cleopatra, Jesus Christ, and the Buddha. Future generations will share the breath of all the signers of the American Constitution, Adolph Hitler, and Marilyn Monroe, and they will also take part in the very breath you are exhaling right now.

This is true of the water we drink, as well. No matter whether it came to you through a faucet or through a bottle of expensive and fancy designer water, or even as part of your pop, coffee, or beer, the process was the same. It fell as rain, flowed as a river, infiltrated into the earth to become ground water. It may have frozen to become arctic ice or cycled through a tree in Tanzania. Once you are finished with it, the water you drink today will once again enter the dynamic water cycle of the earth and circulate around the globe. Even Archie Bunker[70] recognized how interconnected we all are when he said, "You can never buy beer, you just rent it."[71]

So how do you draw a line around yourself? Is the breath in your lungs part of you? What about the oxygen molecules that travel from your lungs into your blood stream? Would you be alive without them? Yet you cannot consider any of them to be "your" oxygen molecules, because when your body is finished with them, they will go, once again, into the soil, the atmosphere, and some will escape out into space to travel throughout the universe. Perhaps, eons from now, the very air you are exhaling today will become a life-giving breath to beings on the other end of the universe. The water (or beer) you drink today will eventually find its way to the ocean, be evaporated into the clouds, fall as rain, and will grow trees, support fish, and make vegetable gardens possible. The heat that your body generates will add to the energy flow of the atmosphere, and the raw materials that you metabolized to create that heat will also be recycled one day. You are not a closed system, but a small bit of the web that connects everything to everything else.

The second observation to discuss is karma. Although karma is often associated with the idea of rebirth into multiple lifetimes, the actual Pali word simply means action, work or deed.[72] So it can also just mean that "you reap what you sow," without having to invoke multiple lifetimes. You don't need to believe that you are the 27th incarnation of King Goomba of Munimula in order to see karma in action. However, there is a philosophical reason that karma is so persistently identified with reincarnation. Since either meaning of karma involves the idea that you get what you deserve, how do you explain children who are born with debilitating illness? Why is there so much inequality, economically, socially, and in matters of health, and why do innocents suffer oppression at the hands of ruthless, inhumane dictators? One solution is to claim that they deserve it because of deeds they did in a past life. But the Buddha's middle way is seen as a path out of this endless suffering, in the life we are living right now.

According to Buddhist tradition, the Buddha himself often said that Karma is intention.[73] The term can refer to individual acts or simply the general principle of cause and effect.[74] In a sense, karma can be thought of as moral interconnection; the actions we perform have an effect on the world around us that continues throughout time, which in turn affects us. We may not be able to know or measure the effects, but we can be certain that they exist.

Let's use an example. Say you own a business and decide one day to belittle an employee in front of the other employees. Not only does that affect your relationship with that employee, but the others now live in fear that you will someday do the same to them. They become more guarded and less able to focus on their work. Quality suffers, and when it does, customers start to fall away. This eventually begins to impact your bottom line, so by one thoughtless action you have negatively affected yourself and those around you in a very tangible way.

We don't need such a dramatic effect to see the results of karma. You may have noticed that the way you treat people affects the way they treat you. Grumpiness begets grumpiness. Smiles grow more smiles. But when you affect the behavior or attitude of those around you, you may never know how this affects their behavior with others, and so on until the actions you take ripple throughout the world. We will be talking later about the butterfly effect in chaos theory, how small actions can cause a chain of events that can result in major effects. If the general idea of karma can be considered equivalent to this concept, personal karma is a sort of butterfly effect from your actions, the butterfly effect of moral causation. The cumulative effect of generations of people making decisions about how to act gives us the world we have today.

Karma demands that we take responsibility for our actions.[75] Yet events are so thoroughly interconnected that cannot always completely understand the full effects of those actions, so we must struggle with our morality in each moment rather than follow hard and fast rules.[76] Taking a rigid and moralistic position on actions does not absolve us of the responsibility of the moment. If an abusive husband asks where the safe house is, will we dogmatically follow our rule not to lie? If a police officer kills an out of control murderer before the killer enters a school, has the officer violated a moral rule not to destroy life?

The world is not static. Every moment, people make new decisions. Every moment rain falls, atoms vibrate, trees absorb sun and water, and planets move through the cosmos. Every moment, though you are not always consciously aware of it, your heart beats, pumping blood through your body and providing life-giving oxygen to your organs. Every moment, you breathe in that oxygen and your lungs transfer it into the blood that the heart pumps. Even systems that we do not consider alive are constantly changing. We look at a chair, for instance, and think of it as a stable and unchanging but if you look deeper, you will see that it is composed of atoms that are constantly in motion. If you consider the effects of time, you will see that the varnish on the chair will yellow and the joints will loosen even in the short time you have it in your home. In the fullness of time, the chair will eventually break down and disintegrate back into the earth.

So another way to describe the dynamic nature of Reality is to say that everything is impermanent. This is the third observation of *Buddha Science*. Everything is in the process of becoming, and this process never ends. The earth is billions of years old. Seemingly tiny changes over long periods of time result in stark and dramatic changes. Rain falls on a mountain, slowly dissolving and transporting the rocks down the slopes. During millions of years, these simple processes slowly melt away the mountain, and it is replaced by a meadow that buzzes with life and change. All is flux.

This is true of everything and everyone. You may know that your body replaces all of its cells every seven years, but did you know that the chemistry of the pancreas changes completely every 24-hours, the intestinal lining is replaced every three days, and 98 percent of the protein in your brain is recycled every month?[77] So what is it, exactly, that makes you the person that you are? An observation that follows from impermanence is that, just like everything else in the universe, you as a person are not static and unchanging.

From this observation arises the concept of "not-self." Since everything within Reality changes and you exist within Reality, it is clear that you undergo constant change. In other words, there is no such thing as an essential, unchanging self. The self that you experience in this moment is a product of all the causes and conditions that led up to this moment, and in the next moment it will change. But the concept of "not-self" goes beyond this.

Try to determine where "you" actually exist. Is it in your arms? If you lost your arms, would you still exist? You can do the same with all of the parts of your body—legs, chest, neck, etc. Most people would draw the line at the brain. I'm dating myself with this, but there was an episode of The Outer Limits, a TV series, where a man is given escape from his ailing body by becoming a brain, connected to two eyes, floating in a vat of fluid.[78] It didn't work out so well for him, but it is interesting because this is pretty much like the image of self that many of us have. We tend to think that "I" somehow exist within my brain.

As we have just seen, the material in the brain is substantially replaced every month, so we can say that the self doesn't reside in the material itself, but what about the content of the brain? If your thoughts and memories could be down-loaded to a computer, would the computer be you? If your body and brain were not destroyed in the downloading process would there then be two of you—one in your body and one in the computer? What if you were cloned? Would there be two of you, or two separate beings? If you were cloned and they sucked out all of your memories and gave them to the clone, then put Homer Simpson's memories into your brain, who would be in your body, and who would be in the body of your clone? This kind of questioning can go on forever, of course. We'll take a closer look at the brain, the mind, and consciousness in Chapter 8.

But what about the soul—is there some intangible, immaterial essence of self that persists forever after the body dies? If this soul is not dependent on physical being then it couldn't have been created when you were born. It must have existed forever. So where was it before you were born? Or if it was somehow created when your physical body was born but persists forever thereafter, does that mean souls can be created but not destroyed? Once again, such questions can continue, and in this case we are venturing into an area that is not encompassed by *Buddha Science*. The question of souls, or Atman, must be left to those of a more philosophical bent. For now, we will say that the answer from *Buddha Science* is that a permanent, unchanging self cannot be found anywhere.[79] The concept of such a self is merely an illusion.[80]

The fourth observation of *Buddha Science* is that the world we think we experience is an illusion. But what does that mean? Beginning from the context of two realities—the ultimate non-dual Reality that encompasses and includes everything and all of time, and the day-to-day reality that we experience—we can see that our day-to-day perspective is limited to observations from the "inside." But from the "inside" observations, we try to build a concept of Reality, often without looking as closely as we should. This results in a belief that the world is a collection of unrelated individual things, beings, and events that may interact with each other but are not necessarily critical to each other's existence. But viewing things as unrelated involves the concept that each contains some unchanging essence that allows them to exist without anything else in the universe. Although there are two schools of Buddhist thought regarding the question of essence, both schools agree that our conventional way of looking at things is an illusion. Individual things, beings, or events cannot be described independent from Reality as a whole.

This ultimate expression of this conventional view is that we, as individuals, are somehow separate from Reality. In other words, there is me, and then there is everything else—"me" and "not me." By unconsciously adopting this view, we feel separated from Reality and we are constantly trying to re-establish this connection. But our efforts at reconnection will not succeed, because we have never lost a connection. There is no connection to lose, because we are an integral part of Reality itself. If Reality is a jelly doughnut, there is no Reality without the jelly. So if it helps, think of yourself as jelly.

Ok, as far as I know they didn't have jelly doughnuts during the time of the Buddha. The point is that *Buddha Science* says if we cling to our belief that we are somehow separate from the rest of Reality, we will never escape the suffering that we create through this clinging. We will never be able to understand ourselves as an integral part of Reality, and so will never be able to think and act accordingly. But it also describes a line of inquiry to follow in order to see Reality more clearly and escape the illusions that cause suffering. Rather than some mystical state of consciousness unattainable except by the most holy, *Buddha Science* lays out a line of inquiry that allows us, even from the "inside" perspective, to see the larger Reality. Like the inhabitants of Flatworld, *Buddha Science* says that by following this line of inquiry we will be able to see beyond our own limited dimension. As we will see in future chapters, in a number of ways this line of inquiry is similar to the methods of formal science.

A WAY OF SEEING

The *Buddha Science* line of inquiry can be summarized as practices that result in a perspective with two parts. These are the practices of developing mindfulness and acting without intention. The resulting perspectives will be called awakening and equanimity. Once again, bear in mind that the perspectives are not some otherworldly trance states but very practical and down-to-earth ways of viewing the world around us, and how it, and we, relate to the larger Reality. Much like the scientific researchers who experience being inside the world of their own inquiry, those who have achieved awakening live in the same world that we do, they simply see it from a different perspective.

The path to achievement of this perspective is rigorous. Where a scientist may prepare for her craft by solving logic problems, mastering calculus, or performing Fourier transforms in her head, a student of Buddhism will work on increasing focus and concentration, practicing mindfulness, and studying at the feet of a master. As with any area of intense study, some students will naturally achieve the shift in perspective in a single, quick flash of insight, while others will require years of work and study before their vision clears. Either way, when the shift comes, it doesn't mean that the work is finished, but that the real work has just begun.

Meditation is the primary method used by the Buddhist to begin on the path. Rather than an escape or trance state, the goal is to train the mind to recognize when concepts are forming in the mind and then let go of them in order to free the mind so that unbiased observation of reality can take place. The most succinct definition and description of mindfulness I have found is the following from *Psychology Today*:

> *Mindfulness is a state of active, open attention on the present. When you're mindful, you observe your thoughts and feelings from a distance, without judging them good or bad. Instead of letting your life pass you by, mindfulness means living in the moment and awakening to experience.*[81]

As this definition implies, the experience of waking up isn't simply an internal experience, it is a way of interacting with the world. In the language of the previous chapter, it means developing a broader understanding of wholeness, and glimpsing the shape of non-dual Reality. Although the complete shape may never emerge, continuing study and practice will allow better understanding.

The next practice to discuss is acting without intention. As the student becomes more mindful of the world around them and how they react to it, they begin to sense when they are leaning in one direction or another. Some things, events, and ideas result in pleasant feelings and others result in unpleasant feelings. This results in a bias toward experiencing the more pleasant feelings, which is the emotional basis of our suffering.[82] We can see how this bias develops as a survival mechanism. Being attracted to sexual pleasure ensures that our species continues to reproduce, and being repulsed by pain ensures that we don't hold our hand over the fire. But when sexual conquest becomes an addiction, or we refuse to run down a burning corridor to escape a building fire, this bias can become dangerous, or even fatal. When the patient walks into the doctor's office, raises his hand above his head and says, "Doc, it hurts when I do this," the doctor replies, "Then don't do that." The doctor's bias is toward keeping the pain at bay rather than fixing the problem, and if he can't see past that bias, the problem may never be fixed.

If the practitioner can learn how to see his leaning, he can separate emotion from observation, so the bias can be detected and his observations become more accurate. In a sense, this is similar to learning how to make objective observations in science, but as we have said, the observations of *Buddha Science* are a subjective view from inside the world rather than the objective view from the outside.

This approach toward observation follows from the discussion regarding two levels of reality. In the non-dual view that is ultimate Reality, there is no subject or object, so the only way to get a glimpse of this Reality is to see through the day-to-day illusion of separation to view every moment through a larger perspective. The apparent paradox in this process is that this perspective can only be fully achieved by acting without intention. Intention implies direction, which means you are presupposing where you want to go, which in turn means you have formed a concept of that destination. If the concept we form makes us uncomfortable, we may lean away from it and lose our connection to the natural flow of life. And as we have discussed, a concept can never be ultimate Reality. In the dual viewpoint, acting without intention may be equivalent to trying to find a destination by driving around randomly instead of following a map. But in the viewpoint of *Buddha Science*, the only way to gain the non-dual perspective is to see it through non-intention.

As an example, the Buddha Scientists of Flatworld might not go about intentionally measuring and calculating. Instead they might observe, for instance, that if one of them sat in one place for a time and other took a journey along one of the very straight paths through their universe, eventually the second one would always come back to the same point from which they started. They might notice that there were shorter and longer paths and if the journeyer took the straight path in any direction that they would always come back to the spot after the same number of footsteps. Although their conclusion would not contain formulas, a Flatworld sage might write something like: "*The longest journey from home and back always takes the same number of steps, no matter the direction.*" Some inhabitants of Flatworld might read this as moral wisdom, others might think it is meaningless drivel, while others would see it as a simple observation that points toward the moon of Flatworld Reality. It all depends on their perspective.

With study, and the practices of mindfulness and non-intention, a student of *Buddha Science* gains a new perspective, one that includes a broader view of non-dual Reality. As we have discussed, any attempt to describe this perspective leads itself to many apparent paradoxes, partly due to the nature of the perspective and partly due to the structure of language itself. In this teaching of *Buddha Science*, we have a subjective view from inside of the object of study, as opposed to the objective view that scientists seek to achieve. In this view, there is no differentiation between subject and object, including the practitioner. As we have said, as part of this perspective, the practitioner himself exists only as a concept, and he cannot be separated from the whole of Reality. This is the perspective that has come to be known as awakening. It is the ability to live in the world and see things from the day-to-day limited view we all share, while also seeing from the perspective of non-dual Reality. It is the ability to constantly be aware of the interconnections, impermanence, and karmic implications of this moment, and the ability to realize that everything that seems so real and solid is, at the level of non-dual Reality, an illusion.

As this perspective settles in and can be maintained through continued practice, equanimity emerges. The term equanimity translates from two separate Pali words that describe different aspects of the English word. The first means "to look over,"[83] and refers to the non-biased observation we have been discussing. The second is a compound word that means "to stand in the middle."[84] So combining the meanings, the word indicates the ability to look at non-dual Reality from the inside without becoming caught up in emotion or clinging to concepts that are at odds with the observations we make. Acting with equanimity is a way to make accurate observations of Reality while being both the subject and object of those observations.

This approach, as we have said, is the *Buddha Science* equivalent of a formal scientist's attempt to make objective observations. The closest example of equanimity in science is a psychiatrist's attempt to stay neutral and avoid becoming emotionally involved, while at the same time becoming intimately familiar with the interior life of a patient.

The subject of psychology, in fact, will be covered toward the end of this book. It is there that we will take a closer look at the heart of *Buddha Science*, the Buddha's observation of the causes of human suffering, and his approach to dealing with the problem. The observations behind the Four Noble Truths, as we will see, are strongly supported by the modern science of evolutionary biology.

This concludes our brief overview of *Buddha Science*. Now we turn to an overview of science, the second finger that we will examine in our quest to point toward the moon of Reality.

CHAPTER FOUR

SCIENCE

The more accurate the map, the more it resembles the territory.
The most accurate map possible would be the territory, and thus
would be perfectly accurate and perfectly useless.
—From the Notebooks of Mr. Ibis: in *American Gods*, by Neil Gaiman

ANOTHER WAY OF SEEING

As the second finger pointing to the moon of Reality, science, like Buddhism, starts from empirical observation. But science takes the position that, in order to be useful, individual experience or observation should be used to develop concepts that can be tested. We'll be discussing some specific assumptions that underlie the scientific method in a few moments, but for now we will just say that there are two basic assumptions of classical science. The first assumption is that we can learn general rules by looking at specific phenomena. For instance, if an apple falls out of a tree and bonks you on the head, it's logical to assume that any time you sit under any apple tree, you could get bonked. If you observe the same thing happening a number of times, you assume that you can derive rules that will always apply, even though you don't measure every possible occurrence of bonking. This might include a rule that each apple drops at the same velocity, the bonk pain is based on the size of the apple, and so forth. The second assumption is that we can separate Reality into smaller chunks and test (or measure) those chunks from the "outside" without affecting the results. So, for instance, let's say you are trying to determine the likelihood of getting bonked when you sit under an apple tree. After a number of trips to the emergency room with concussions and the associated loss of data, you decide to hire an assistant and conduct your measurements while they are getting bonked, on the assumption that the same bonking rules apply to your assistant and will not be affected by your measurements. In this case, you can be pretty confident in your assumption. As we will see later, things aren't always so straightforward.

Based on these two assumptions—that a limited number of tests can be used to derive a general rule, and that you can separate parts of Reality and test them without affecting the result—scientists utilize a specific methodology in order to progressively improve the state of scientific knowledge. This is the heart of the scientific method, which can be outlined as follows:

- Make an observation

- Ask a question about it

- Conduct research to find out if others have provided answers

- Form a concept (hypothesis) to explain it

- Design and conduct an experiment to test the concept

- Analyze the results

- If the results do not match your expectations, come up with another concept to test

- If the results do match expectations, try more tests to prove or disprove the concept

- If tests continue to support the concept, accept it as true

- Any time new tests disprove the concept, modify or reject it and begin testing again

In this way, science relies on inductive reasoning to develop general rules to describe Reality. Inductive reasoning says it is valid to use a limited number of observations to develop general rules for how things are. Inherent in the inductive method is the development of concepts that can be disproved, so that if anyone can find instances where the concept does not apply, either the concept or the observations may be thrown into question. At this point, scientists must decide which of the two is more credible, and either reject the observations that seem to disprove the concept or develop another concept that incorporates the results. It also may turn out that new testing reveals a special set of circumstances under which the original concept doesn't apply, while in other situations it remains valid. An entire set of criteria can be established to help scientists decide how much evidence is necessary to disprove observations, which seem to contradict well-established concepts.[85]

The idea that concepts can be disproved by testing can be called falsifiability, a concept made popular by Karl Popper, a philosopher and professor who lived from 1902 through 1994. Basically, his position was that the inductive reasoning utilized by science means that scientific theories can never be proven, but they can be shown to be false. This raises a couple of issues. First, if something can't be proven false, you can't study it in a scientific way. For instance, I could reveal that in a past life I lived on the planet Xenon in the Sofabed galaxy, and a scientist could not disprove my claim unless he located the planet and invented a time-traveling spaceship to get there during that past life. The second issue is that any new discovery can dramatically shift the direction of science if it demonstrably contradicts existing scientific thought. The state of scientific knowledge at any point in time is based on concepts that fit the existing information. For instance, if a scientist found a massive vacuum cleaner at the center of the earth, and everything started floating when she switched it off, she might have to rethink the idea of gravity. Like most of us, scientists typically begin from the assumption that well-established concepts are true until proven false.

This is a major point of departure from *Buddha Science*, which states that everything that exists now, and everything that ever was, is interconnected. Therefore, everything is so complexly related that it is impossible to completely understand, intellectually, how it all fits together. Inherently, says *Buddha Science*, we observe this truth, but our minds get in the way by trying to form a concept around it. Concepts, in this view, are the hobgoblins that get in the way of accurate perception. In other words, you can perceive non-dual Reality only through getting rid of all conceptions of it. This is pretty much the opposite approach to that of science and its reliance on concepts to discover Reality through inductive reasoning.

Science does agree with *Buddha Science* that to begin by trying to understanding all of Reality at once is bound to failure. But scientists are of different minds when it comes to addressing this problem. Reductionist science says that Reality can be conceptually understood by studying the pieces and building upon the findings to ultimately combine them into an accurate, but useful, map. Holism, on the other hand, says you must try to understand the big picture first, and this will tell you how the pieces fit together. In practice, science generally settles on an approach somewhere between the extremes of reductionism and holism.

ASSEMBLING THE PIECES

Historically, the progress of scientific discovery is a lot like people putting together a jigsaw puzzle. There are two "extreme" approaches to jigsaw puzzling. You can look at the picture and, based on the colors and textures in the picture, try to decide where each piece fits within that picture. Or you can look at the shape and color of each piece and try to see how it fits with the adjacent pieces. I don't know anyone who closes their eyes to avoid looking at the picture that they are trying to put together. But I also don't know anyone who uses the picture without also looking at the shape and color of individual pieces. Generally, it's a process of working from both ends.

Science certainly points to interconnectivity, but does it through the development, application, and testing of concepts using the scientific method. Science says that although things are interconnected, we can pull out parts of the system, form a concept of how each part works, and observe that part of the system objectively from "outside" that part without significantly affecting the results. Then we take the bits, compare them with our current concept of Reality, and see how they fit into the puzzle.

In this way, concepts are critical to the scientific method. Models that are built on these concepts, however, are only as good as the level of detail that goes into building them. In the next chapter we will see how chaos theory has demonstrated that even simple models can exhibit wildly different results through "sensitive dependence on initial conditions." In the view of science, correct concepts, theories, and models are critical to making sense of our observations, and therefore to making them useful. A key part of the scientific method, however, is to make certain that we don't confuse our concepts or models with Reality, because we must be willing to modify those concepts in light of new information that proves them false. This is another place where the methods of science and *Buddha Science* agree.

In the way of illustration, imagine you follow a map to a state park, but when you get to the location, there's a poultry farm where the park is supposed to be. You may assume you have not followed the map correctly, or that something is wrong with the map itself. In any case, it's clear that the map and the reality are not the same thing. But if the map has been right up to this point, would you assume that it's a mistake to use a map at all and just throw it away? More likely, you would rationalize the error by thinking that perhaps the park was closed down and sold to the poultry farmer after the map was made. After all, it was one little glitch. But if the map is somehow flawed and you continue to rely on it until enough errors crop up that you become hopelessly lost, at what point would you give up and buy a GPS unit? Of course, if the map is badly wrong you won't be able to use it to find a store to buy the GPS. Assuming you eventually do find a store and make your purchase, if the farmer did buy the land, the GPS could be wrong if the software hadn't been updated. But having some navigation tool is still more useful than wandering around trying to find the park by pure instinct, at least according to a scientist. If you were a Buddhist practicing non-intention, perhaps you would simply enjoy the journey.

On the other hand, imagine that you found a dried up peach that seems to resemble the terrain, and the first couple of times you used it, you successfully found locations you were traveling to. Since then it's been wrong 48 times, but it's still better than just driving around randomly, right? If you really believe that, you might want to rethink your approach rather than continuing to cling to the peach concept of navigation. Either way, it is important to keep the goal in mind. If the thing we are interested in is the actual location of the state park, then the map, GPS, or peach are only fingers pointing to that location. None of them is the territory itself.

These examples are contrived, of course. I would never admit to using the dried peach method of navigation. But they illustrate a dilemma that scientists on the cutting edge of their profession deal with on a regular basis. Because the scientific method is based on using inductive reasoning to form concepts that apply universally, there is always the possibility that the concept may not apply in any given situation. If an experiment seems to dispute your pet theory, did the experimenter that found the problem follow the correct procedure (or correctly follow the map)? Are there limited cases when the concept doesn't apply (such as the property being sold to a poultry farmer)? Is a new, more comprehensive concept needed (buy a GPS)? Is the concept simply wrong, and you have been lucky until now (dried peach navigation)? Science relies on a complex structure of rules, assumptions, and experiments, so every time a new discovery appears to contradict established wisdom, the scientific community struggles to readjust. As fresh concepts emerge to incorporate the new information, science progresses.

The basic idea of progress in science, as developed through the process of inductive reasoning, is that we can discover Reality thorough an iterative process—observing, proposing, and testing ideas, and preserving those that seem to point to Reality as scientific theories or laws. In this process, demonstrated and repeatable observations are the only "facts" in science. Theories and laws can be modified or rejected at any point if new observations come to light, or if new concepts are proposed that seem to more accurately describe the collective observations of the scientific community. In this way, the goal of science is to progressively correct and refine its conclusions and to come closer and closer to describing Reality in a way that can be understood and described, and therefore is useful.

So beyond the basic two assumptions we have already discussed, what are some other assumptions that go into this approach? There are many views, and lists of assumptions, which in itself is testimony to the view that scientific "facts" themselves can be subject to question. There are literally hundreds of lists that try to describe the assumptions that support the scientific approach. The list below is an attempt to capture some of the ideas, and the selection of what to include is primarily based on my experience as a scientist. The list itself is only a map, not the territory. Some abbreviated discussion is provided here that I hope will prove useful. We will be elaborating on various points in upcoming chapters.

So here is my list of assumptions that are the basis of discovering Reality through scientific methods:

1. There really is a world external to us that exists, with or without us to observe it.[86]

2. Reality is rational and orderly, and follows rules that we can discover and describe.[87]

3. Everything follows these rules, even those things that we can't yet see or measure.

4. We can measure things objectively, i.e., we can step outside of what we are measuring and measure it without affecting it or Reality itself.[88]

5. Pure science and logic will drive the process of discovery, so that any findings contrary to the established assumptions will be greeted with objectivity, embraced, and thoroughly investigated.[89]

The first assumption, that there is a world "out there," is based on the dualistic concept that there is an "us" and a "not us"—i.e., "not us" is the rest of Reality. Trees fall in the forest whether or not someone is there to see it happen. The moon exists, even when we cannot see it. Therefore, there is an observer and something that is being observed. Even though everything may be connected, we can ignore the connection between the observer and the observed at most scales of measurement. Careful consideration of this question quickly leads to questions of consciousness, a subject that cannot be explored without finding contradiction and paradox.[90] After all, consciousness is a subjective personal experience that does not yield to consensus.[91] In fact, exploration of this question can quickly lead to the question of whether it is possible to ever be objective about anything.[92] We will be exploring questions of consciousness in Chapter 8. However, this first assumption is so basic to science that any other view falls into the realm of metaphysics. So from the viewpoint of science, we will accept the idea that Reality exists, whether or not we are around to observe it. The map is not the terrain, and science itself is only a map that cannot be useful if some assumptions are not allowed.

The second assumption, that Reality is rational and orderly, is ingrained in traditional science, and most specifically in the Newtonian view of the universe as a perfectly functioning clockwork, fully deterministic and predictable if we could only gain enough knowledge to know all of the governing laws. As part of this assumption, if experimental measurements show slight variation, the experimenter generally assumes that the variation is due to the limitations of measurement itself rather than any fundamental variation in the rule that governs what we are measuring.[93] Although the discovery and elaboration of chaos has modified this view by showing that some behaviors are not predictable, it also demonstrates that there are broad patterns that can be quite predictable. These will be explored further in the chapter on chaos. We have already discussed the limitations of language, which confines our ability to adequately describe the rules of Reality.

I have added the third assumption, that everything follows the rules. This is from the inductive method—the idea that we can discover rules by measuring a limited number of cases, and assume these rules will hold true for all other situations. Many phenomena exist that currently cannot be measured or fully explained, but new discoveries and techniques may find a way to investigate them further. A prime example of a discovery made possible through technological advancement is the discovery of the Higgs Boson.[94] This particle could never have been directly discovered without building the Large Hadron Collider. The Higgs, or perhaps some other form of elementary particle, may lead the way to revealing the mysteries of dark matter and dark energy, which is currently invisible to us even though it constitutes 96 percent of the universe.[95] So even though the unknown universe is 24 times as abundant as the universe we can know and measure, we still assume that we can determine the rules of the universe without fully understanding the phenomenon of dark matter and dark energy. Could the discovery of the Higgs lead us to develop new ways of measuring that will allow us to evaluate this phenomenon? Could this bring information that will cause science to re-evaluate everything we think we know about the rules of Reality?

The fourth assumption is closely related to the first. Can we really measure things objectively without affecting the result? The question not only relates to the ability to step outside the system and measure a phenomenon objectively, but also to the issue of whether the measurement itself affects the outcome of what is being measured. In the example of Flatworld, the inhabitants were able to gain insight into a broader view from the inside and were also able to effectively measure their world without affecting it. But as physicists attempt to measure smaller and smaller phenomenon, they are discovering that the observations cannot be separated from the act of observing.[96] In fact, it is also possible that any actions, however small, may be so affected by all other actions in the universe that the precise effect of observation cannot be calculated.[97] The question remains: If there is some effect from every action, at what level of measurement is the effect insignificant enough to the outcome that we can ignore it?

The fifth and final assumption listed above doesn't have to do with the nature of the Reality we are trying to measure, or our capability to measure it. Rather, it is the assumption that the scientific community has the ability and will to follow the methodology that it has set for itself. Although this process is well defined in the physical sciences, applying the scientific method can be more problematic in the branches of science dealing with the mind, or with life processes. In addition, such subjective influences as publication bias, and the potential tendency of scientists to unconsciously steer their results toward support of established assumptions, can make it difficult to perfectly apply the scientific method. This potential bias may be entirely subconscious. Indeed, beliefs based on established assumptions may be so thoroughly ingrained that they form a sort of blindness, making it virtually impossible for a researcher to accept results that do not support those beliefs.[98] If the scientific community can't accept the results because they shake up the status quo, revolutionary discoveries can be, and have been, ignored for years. Thomas Kuhn provides a great amount of insight on how the scientific community deals with revolutionary discoveries in his landmark book, *The Structure of Scientific Revolutions*.[99]

Experimental design is also related to this final assumption. Typically, experiments are designed to test a particular concept, so an experiment may potentially introduce unexpected bias into the results. Is it possible for scientific inquiry to be completely objective when experiments are designed around the scientific view of Reality that is popular at the time? After all, if a scientist thought an illness was genetic, they might design a very different test than if they thought it was caused by, say, evil spirits. Although test design tends to be more of a concern with the social sciences, an ongoing problem in quantum physics is the observation that experiments often find exactly what the researcher was looking for, raising the question of whether experimental design actually creates the expected result. On the other hand, many significant discoveries have resulted from lab mistakes, accidents, or flashes of insight seemingly unrelated to intentional experimentation. Yet it is impractical and naive to expect experimental scientists to poke about randomly in the hopes of making the next earth-shaking accidental breakthrough, just as driving around randomly to find specific locations will not yield reliable results.

There's a sixth assumption that is included in some lists. This is the assumption that mathematics effectively describes Reality. I haven't included it here because we are surveying various types of science, and the application of and reliance on mathematics varies significantly depending on the branch of science that is being studied. Although some would consider mathematics to be the science of numbers, scientists are divided on whether to consider mathematics as a map that can be used to negotiate the terrain of Reality, or a description of Reality itself.

Is 42 the answer to everything? Is mathematics purely an intellectual construction of human thought, or is it a law of the universe that is only slowly being discovered as we encounter new situations through scientific investigation? There are those who would come down on each side of this question.[100] Yet when a university student is being schooled in the various scientific disciplines, he is required to internalize various mathematical concepts in order to graduate. From a practical standpoint, this is necessary because scientific thought is often described in the language of mathematics, and this language makes innovation and invention possible. However, after wrestling with and mastering the difficult intellectual concepts associated with learning calculus, statistics, or multi-dimensional geometry, such concepts begin to take on the cast of Reality. Yet most differential equations cannot be solved at all,[101] and we have already seen that the circle, one of the simplest geometric forms, cannot be completely known from a mathematical perspective. So if mathematics is actually a window into Reality, then we understand very little about Reality. Yet, through the educational process, scientists often enter their profession with a built-in bias toward a "belief" in mathematics. As we discuss the various branches of scientific study, we will see that the influence of mathematics comes up often.

The next chapters of this book are organized around these various branches of science. We have divided the discussion into four broad categories:

- Chaos and complexity, which includes considerations and principles that can be applied to all of the other categories.

- Physical sciences, primarily relativity and quantum physics

- Life sciences, which include studies such as biology and genetics.

- Science of the mind, primarily including psychology, neurology, and consciousness studies

As we explore each one of these categories, we'll be looking at specific details and findings within that category, and the parallels between those findings and the observations of *Buddha Science*. In addition, to the extent that the categories share specific findings, we will bring that out as we go along. Once again, although we have tried to cover each category in sufficient detail to provide an in-depth discussion, the topics are much broader in scope than is possible for this book to explore fully. Hopefully, the references will provide a useful starting point for you to delve deeper into topics that may catch your interest.

CHAPTER FIVE
CHAOS AND COMPLEXITY

"I'm writing a book on magic," I explain, and I'm asked. "Real magic?" By real magic, people mean miracles, magical acts, and supernatural powers. "No," I answer: "Conjuring tricks, not real magic." Real magic, in other words, refers to the magic that is not real, while the magic that is real, that can actually be done, is not real magic.
—*Net of Magic*, by Lee Siegel

AT THE MOVIES

Chaos scares and fascinates us. Alfred Hitchcock understood that when he filmed *The Birds*. Start with a beautiful little town in northern California. Then one odd thing happens: a gull hits someone in the head, creating a bloody gash. Was it a fluke, an accident? Or was it something more sinister? Then slowly the birds begin to mass. One sits on a wire, then two, then eight, then more and more of them. They take flight, swirling angrily in the air. Their attacks become bolder and bolder, and bit by bit it becomes clear they are massing for a full-out assault on the residents of the town. Chaos ensues. Once we see the pattern in horror movies, we can see it again and again—*The Shining, Jaws, Halloween, Poltergeist*—the list goes on and on. Many doomsday action movies have the same patterns. I actually connected with my wife through movie chaos. On our first date, I took her to the Tim Burton movie *Mars Attacks*, which takes chaos to ridiculous extremes. The fact that she actually enjoyed the movie gave me hope that she would put up with me for the long term.

Think about your own life experience. When things get chaotic, how do you react? Even if you "thrive on chaos" as some popular books advocate, you probably feel a little bit crazy when things spin out of control. People have different tolerance levels for control, or course. Some people need adrenaline-pumping sports to feel as if they can control their mind and body, and some people get flipped out when their cherry jelly is the wrong shade of red. The chaotic climate of war or extreme violence causes people to live in nightmare for years after the actual events. But whatever level of chaos it takes, you can probably identify a very specific threshold of activity when life starts to make you crazy. You step through the doorway from a sane, orderly world into a world where everything is out of control. When you cross this threshold, you experience a feeling of chaos.

Chaos surrounds us, although it is generally less disastrous than the movies, or our worst fears, would suggest. The scientists who first discovered the principles of chaos theory must have shared a similar discovery process with our hapless movie victims: seeing unexplained patterns, recording seemingly simple data with a predictable trajectory that split, then split again, and suddenly and unexpectedly broke into chaos. It must have seemed like real magic. Once scientists learned to see the patterns of chaos, the patterns began to show up everywhere.

Through the movies, and through our life experiences, we have learned that chaos is a scary thing. We have nightmares of swirling dark waters, of tornados and swarms of angry, buzzing insects. Heart attacks are related to chaos in the electrical impulses coursing through our bodies. But chaos is a pleasant part of our everyday life as well. It is in the water boiling for a calming evening tea, in the motion of the wind chimes that let us know a storm is on the way, in beautiful winter woods and rugged majestic mountains. Scientists have found that chaos is virtually everywhere, connecting things in astonishing and unpredictable ways.

But chaos is only the beginning. The discoveries of chaotic behavior in simple systems came along at the same time that physicists were struggling to link relativity and quantum theory. Scientific thought has evolved rapidly in the last couple of decades based on the recognition that systems cannot be reduced to a collection of component parts. Two interrelated lines of inquiry have emerged, those of chaos and complexity. Each of these lines of inquiry is an overlay to a wide range of scientific disciplines. In other words, the principles involved in these lines of inquiry can be applied to many areas of scientific study.

MOVIES, SCHMOVIES

We begin with chaos. Ok, I started out with the movie-theater scary kind of chaos, but now we're going to get into the scientific kind. Although the two are related, since scientists are—well, scientists—you might imagine that they have a pretty much standard definition of chaos. It might take the scary fun out of it, but it's still pretty fascinating stuff. We can wait while you finish your popcorn.

From a scientific perspective, chaos can be described as the observation that simple systems that contain a feedback mechanism can exhibit unpredictable behavior. As the science of chaos has evolved, its definition has narrowed. In fact, as the study of complexity has gained traction, chaos has come to be defined as a very specific type of behavior that contributes to the complexity of systems.[102] The study of complexity seeks to understand the functioning of whole systems that have many component parts. It seeks to discover and describe how these parts interact, and how the behavior of the system relates to its parts. These efforts lead to a third subject, that of emergence. Emergence is simply the observation that the behavior of a system is more than merely the sum of its parts. In this sense, it is a rejection of reductionism, which as we have described, says that you can break a system into parts, study each part, and then put it together to explain the functioning of the system. The principle of emergence says that at some critical level, systems demonstrate an emergent, system-wide behavior that cannot be predicted or anticipated by simply studying the parts. So the three general topics of interest in this chapter are chaos, complexity, and emergence. Let us now consider each of these three topics in more detail.

Chaos is one of those terms, like Buddhism, that has been applied to a wide range of phenomena. In current parlance, the term chaos is limited to the observation that small changes in the initial condition of simple feedback systems can have dramatic effects. Such observations have been with us since ancient times, and specific instances have been described in the scientific literature as early as the 1860s.[103] The idea of chaos as a scientific concept worked its way into the popular imagination after a meteorologist named Edward Lorenz tried to construct a mathematical model of the weather; he found that if the initial conditions varied only slightly, the modeled weather changed dramatically within a few (simulated) months.[104]

Lorenz called this "sensitive dependence on initial conditions," but it became known as "the butterfly effect" after Lorenz failed to provide a title for a paper he was presenting at the American Association for the Advancement of Science in 1972. According to Lorenz, the person responsible for compiling the program book concocted the title, "Does the flap of a butterfly's wings in Brazil set off a tornado in Texas?"[105] This "butterfly effect" so captured the imagination of the audience that it entered into the general lexicon. It is now a concept almost as familiar as our rock star number, pi. I will grudgingly admit the current widespread use of the term "butterfly effect" is probably due in part to a rather disappointing movie that came out in 2004 about a guy trying to fix his present by traveling into the past. Sort of like a "Ground Hog Day" without Bill Murray or decent humor. But in true Hollywood fashion, it spawned its own butterfly effect with two sequels that were even worse.

Karma is the *Buddha Science* moral analog of the butterfly effect. Since human interaction is so complexly interrelated, you can ask if smiling at the grocery checkout clerk in Chattanooga can prevent a war in Uganda. Call it the social butterfly effect.

Although climate scientists generally agree that an effect as small as a butterfly wing would probably get damped out by the atmosphere, small changes in a feedback system can become amplified to create dramatic effects. One clear way to demonstrate this general principle is to turn up the microphone just a little too high on a public address system. As you turn up the volume on the system, it behaves in a very predictable way as the sound becomes progressively louder. But once you reach a particular threshold the system emits an earsplitting squeal—the onset of chaos. If you set the speakers behind the microphone, the squeal will begin at a much lower volume setting than if you move the speakers forward where the microphone picks up less sound from the speakers. You can also use a directional microphone that picks up less sound scatter from the speakers. It is the feedback component in the way you set up the system, the initial conditions, that determine the volume level needed to initiate the sound of chaos. In a systems sense, this follows from the laws of cause and effect.

But chaos demonstrates that even if we know the fine details of a system, we cannot necessarily predict its behavior. Chaos refutes the reductionist idea that we can understand the whole by discovering everything about its parts.[106] As we have discussed, reductionist science is based on the proposition that the universe works like a predictable machine, so we can understand the workings of the machine by examining the parts. If we could measure everything perfectly, we would be able to perfectly predict everything that is going to happen. But noise gets in the way of perfect measurements, so the ongoing quest is to look for the signal by progressively perfecting our measurements in order to eliminate all of the noise. Chaos turns this idea on its ear, by saying that the noise is the signal. The very object of study is that noisy, unpredictable part of the system that a reductionist wants to throw away, or ignore. The same public address system will break over into the "chaos squeal" at different volume settings, based not only on the system settings, but the size and shape of the room, the room furnishings, indoor versus outdoor settings, and even the atmospheric conditions. Rather than turning the volume down as soon as a squeal is produced, the chaos scientist might look at the amplitude and phase characteristics of the sound and study how those characteristics change with system settings and other conditions to see how all of the various factors affect the results.

There is a clear intersection here with the observations of *Buddha Science*. Since everything is interconnected, it follows that every action has an effect on everything else, and the butterfly effect is simply based on the cascading effects of a complex sequence of small actions. Each action is unique, and the result is dependent on so many actions and relationships that prediction is impossible. The principle of karma also suggests that human effects continue indefinitely in ways that we can never understand, a result equivalent to Lorenz' observation that long-term weather prediction is impossible because we can never fully understand and model all of the inputs that effect such a system. So we are not only intimately connected to our past, our actions also significantly affect the future, no matter how small those actions may seem today.

Lorenz also asked if the weather would settle into an equilibrium state and concluded that it never would. The idea of a long-term average is meaningless, since a stable, unchanging climate does not exist.[107] Perhaps the weather was invented in Lake Wobegon, where averages don't apply. Though modern weather reports average historic temperatures as a guideline, such comparisons are an illusion, a concept that cannot really predict what the weather is going to do tomorrow, this week, or this month. In the words of *Buddha Science*, the weather is impermanent, a process instead of a thing, never the same from moment to moment. If you have any illusions about weather predictability, try living in Michigan.

The current concerns about climate change are warranted because of the direction that change is taking, even though climate itself is defined by change. It's actually unfortunate that scientists settled on the term "climate change" for this very reason. The problem is not that the climate is changing, but that it is becoming more predictable in the long term and less predictable in the short term. The longer-term trend is in the direction of warming, which has significant ramifications for the entire world population and economic systems. In the short term, this translates to more unpredictable weather patterns as the entire system adjusts to find a new dynamic. Although I took a longer time than most scientists to become convinced, the recent unprecedented storms, droughts, bird migration patterns, changes in the ice sheets, and many other independent lines of evidence present compelling support that we are in a dramatic and long-term warming trend.

Unpredictable behavior has been observed in many systems studied by scientists. Fluid turbulence cannot be fully modeled or measured, even with the best computers and measuring instruments we have today. Although earthquake distribution patterns can be mapped, no tool exists to accurately predict major quakes in a consistent way. So-called empty space changes so dynamically that we can't even see the virtual particles that pop in and out of existence.[108] Even the seemingly simple problem of attempting to predict the next drip of a faucet can be elusive. As Niels Bohr once observed, prediction is very difficult, especially when it's about the future.

Yet out of seeming randomness, order can emerge. Each snowflake is unique, based on the particular path it takes through the atmosphere and the conditions such as temperature, humidity, and impurities it encounters. But even though each individual snowflake is different from each every other snowflake, each develops a near perfect six-fold symmetry.[109] So each snowflake will have six segments radiating around the center, and each of these segments will look more or less identical to the other segments. Snowmaking creates endless, unpredictable diversity, but very predictable symmetry. We will also see that many systems show chaotic behavior interspersed with periods of order.

One way scientists map a larger order within a chaotic system is to translate it into what is known as a phase plot. This is a diagram that shows all possible states of a system. For instance, a scientist might plot physical location against momentum to try and understand the motion of a pendulum. Or a stock analyst might plot the price of a stock against time to try and predict whether it is going to go up or down. A mathematician might plot real numbers against imaginary numbers to understand the possible outcomes of a set of equations. Such a plot is a way to make a visual map of numbers, showing patterns that may not be otherwise observable. If the plot shows a particular shape toward which the system evolves, the shape is called an attractor. Once again, the scientific term has a very specific meaning. So in scientific lingo, an attractor would not necessarily refer to a person who attracts others in a bar.

The simplest type of attractor is a point attractor. As an example, take a simple pendulum—a plumb bob on a string. If you pull the plumb bob off to the side and let it go, it will swing around for a while, slowly settling down until the string hangs vertically. This is a point attractor. It will stop at the point in phase space where the velocity is zero and the position is directly below the place where the string is attached. If the plumb bob is metallic and you put a magnet to one side, the plumb bob will still settle into one position, but will be deflected toward the magnet so the string will not be perfectly vertical. This is also a point attractor, but the point has simply moved to a different place.

By comparison, a pendulum that is mechanically driven, such as the one in a grandfather clock, will continuously repeat the same action, resulting in a cyclic attractor. This type of attractor will show a distinct shape, which may be a simple circle or ellipse, or it may be more complicated if the system takes more than one cycle to repeat itself.

By contrast, a chaotic system will theoretically result in a pattern that never precisely repeats itself even though the shape of all the possible configurations can be plotted. The shape of snowflakes we discussed above could be modeled as a strange attractor due to its endless variety around the pattern of six-fold symmetry. The pattern that forms in this type of system is called a strange attractor—strange because although the pattern itself can be known, the path that it takes never exactly repeats itself. As we saw with "irrational" numbers, it's interesting to note that if something doesn't fall into a satisfying and predictable behavior, science labels it as "strange."

Studying attractors is a way to step outside the system and see what the overall shape looks like, even though you may not be able to predict the exact path the system will take. It's a way to see the elephant. Consistent with *Buddha Science*, chaos shows us two realities—one that is concrete and measurable, and ultimate Reality, which can be intuited but not fully understood. In the language of non-duality, chaos does not exist without order, and order does not exist without chaos. A system in chaos has a predictable shape, even if we can't predict exactly what the state of the system will be in any given moment. The tools of chaos have given scientists ways to study both of these aspects of the same system.

So the concept of a strange attractor is basically a rigorous way to describe a system that never exactly repeats itself. In such a system, each moment is unique and unpredictable, even though you can completely describe the limits on the system. Examples of these systems abound. One example is the breathing process. Breathing is a continuous looping system with upper and lower limits (lung capacity on the upper end and the capacity to expel air on the lower end). Within those limits, the length and velocity of each in and out breath can vary, and the length of the pause between breaths also varies. One primary focus of meditation is following the breath as it changes, noting that each of these factors change constantly, that each breath is different and unrepeatable—a perfect description of a non-linear system that could be illustrated as a strange attractor.

You can experience this process for yourself, simply by setting aside some time when you will have no distractions and finding a quiet place to sit. Simply focus your attention on your breath, how it flows in and out. Don't try to control it, just observe your breathing as it happens. You will find that some breaths are short, some longer. Some are deeper and some are shallower. The velocity of in-breaths and out-breaths change, and so does the transition between in and out breaths. There is a quiet moment between breaths when everything pauses. This moment can be a time of incredible calm, and as with the other aspects of breathing, this moment varies in length as well. With continued practice, you will find that each breath, and each moment of each breath, is unrepeatable. This is a direct Buddha science experience of chaos. You could say that chaos is a scientific way of saying that every moment—every "now"—is unique.

But breathing can not only be experienced. Using the tools of chaos, it can also be modeled, and the results can be applied in an effort to improve lives. In fact, in 2008, a researcher in Australia conducted a study aimed at using the mathematics of chaos theory to predict the breathing of infants during sleep, in order to effectively monitor for problems.[110]

Scientists use mathematical sets as maps to help them understand chaotic behavior. These mathematical sets are developed by creating an equation that feeds back on itself. The scientist feeds initial values into the equation to begin the process and the equation spits out a result. The resulting number is fed back into the same equation to calculate the next value, and this process continues like the microphone on the public address system that feeds back the output sound from the speakers. This continues over and over again and a plot is created. If the plot has elements that repeat themselves on smaller and smaller scales, the shape is said to be fractal. This repetition of pattern at smaller and smaller scales is called self-similarity, and can refer to patterns that repeat exactly, or to patterns that are very similar. Probably the most famous of these plots is called the Mandelbrot Set, named for Benoit Mandelbrot, the scientist who discovered it. This set is plotted on a special map that has real numbers on one axis and complex numbers on the other. The boundary between the numbers that are within this set and those that fall outside the set is so complicated that the set has been called the most complex object in mathematics.[111] The most fascinating and counter-intuitive feature of this boundary is that the boundary itself is infinite in length, even though it encompasses a finite space in the complex plane. Although it was first described as a pattern by itself, the Mandelbrot set has shown up in other places as well. For example, in 1986, researchers who were studying the phase transition between magnetized and non-magnetized materials published photographs that documented an unexpected result. They were startled to find the Mandelbrot set deep within their data. Their reaction? "Perhaps we should believe in magic."[112] To paraphrase Bullwinkle, "Wanna see me pull a Mandlebrot set out of my hat?"

The very concept of fractals seems like magic. Many systems exhibit patterns that repeat themselves at smaller and smaller scales, which results in many questions about our concepts of space and dimension. For instance, how do you measure the circumference of a cloud? Do you simply put a big circle around it, or do you try to measure every twist and turn of the cloud's boundary? After all, the boundary of a cloud looks pretty similar close up as it does from miles away. Do you try to measure the boundary down to an individual droplet of water, to the molecular level, the atomic level, or even father? As you measure it at smaller and smaller scales, the boundary becomes more complex and therefore longer. Just like the boundary around the Mandelbrot set, if you could measure a cloud with an infinitely small ruler, the boundary itself would become infinitely large.

How can this be? Aside from Donald Trump's ego, how can something infinite be contained in finite space? One other way to think about fractal objects is through the concept of effective dimensions.[113] Instead of viewing the cloud as a complex object existing in three-dimensional space, it could be viewed as an object that exists in a dimension somewhere between two and three. In other words, the object itself defines the space, rather than the space defining the object. One way to interpret this concept is that objects don't fully exist in two or three-dimensional space, but in some fractional space specifically defined by each object.

Wait a minute! Have I been smoking funny stuff again? How can the space define the object? Isn't the world comprised of discrete objects separated by space? Once again, this is a perspective that is difficult to shake, but it is only a concept. Are our objects defined by their shape, or the shape of the space associated with them? Consider the following from the Tao Te Ching:

> *Thirty spokes share the wheel's hub;*
> *It is the center hole that makes it useful.*
> *Shape the clay into a vessel;*
> *It is the space within that makes it useful.*
> *Cut the doors and windows for a room;*
> *It is the holes that make it useful.*
> *Therefore profit comes from what is there;*
> *Usefulness comes from what is not there.*[114]

As another example to illustrate the concept of fractal space, consider a perfectly flat and perfectly thin piece of paper. Ignoring the minimal thickness, we could say it exists in two dimensions. But what happens if we warp it or fold it? What happens if we crumple it into a ball? It is still a two dimensional object? What if we could take that ball and perfectly compress it into a ball without any space between the crumples—would it then magically become a three-dimensional object? Have we changed the object, or the space around it? Scientists use the idea of fractal dimension to describe this. As the paper is crumpled into a smaller and smaller ball, the folds become more complex and the fractal dimension increases incrementally between two and three dimensions. Space and dimensionality are simply concepts based on relative measure, dependent on the idea that individual objects exist as something separate from the space that surrounds them. The concept of fractal dimensions is another way of looking at the relationship between objects and that space. As a Buddha Scientist would say, these concepts have no meaning from the perspective of non-dual Reality.

COMPLEX BUT NOT CHAOTIC

You might observe that the ball of paper exhibits self-similarity because the smaller folds look a lot like the larger folds, except they are—uh, smaller. But not all complex objects, or systems, exhibit self-similarity. A system might have some self-similar elements and some that are not structured in the same way. The human body is a prime example. The bronchial tubes in your lungs show self-similarity, branching into smaller and smaller units until they become bronchioles, which are the smallest subdivision of this structure. The bronchioles are attached to air sacs, where the oxygen you breathe is exchanged for the carbon dioxide that your system needs to get rid of. This complex structure allows for the maximum number of air sacs possible in the space allowed, thereby maximizing this exchange of gasses. The heart is also part of the circulation system, but looks and functions in a much different way than the lungs. Internally, the heart consists of muscles, valves, veins, and arteries, so while it has some elements that are similar to each other, many of the parts look very different from each other. So the heart organ as a system cannot be considered self-similar. The same, of course, is true of the human body as a whole, with its various systems, organs, bits, and pieces.

Yet the human body functions as an integrated system. All of the various parts work together for the benefit, or failure, of the whole. This is a complex system with some chaotic elements, but it cannot be considered chaotic as a whole. The study of such systems has become a focus for the scientific study known as complexity theory. This field cuts across all areas of scientific study, including systems as far ranging as physical systems, social and political systems, and intelligence studies. The main focus of these studies is to determine how the parts of a system relate to the whole system, and how the system itself interacts with its environment.[115] The parts themselves may be simple or complex, and there may be no chaotic elements involved. The key is that these parts combine to form a larger system that functions as a whole.

But the whole is much more than the sum of its parts. In Mary Shelly's novel from the 1800s, Victor Frankenstein discovered that it's hard to build a human being out of spare parts. Not all such discoveries involve such drama, but they may be equally enlightening. Take water for instance. A molecule of water is not "wet." It takes a number of molecules, acting together, to possess the quality of being wet.[116] This quality is termed "emergent," because it emerges as a system reaches a certain level of complexity. An emergent behavior of the system as a whole cannot be found or generated by any of its parts acting alone, or even a few of the parts acting together. The emergent behavior is specific to the complex interactions of all the parts acting together. It is therefore not simply the sum of the parts, and can only be understood by studying the system as a whole. In other words, the emergent behavior is indivisible from the system as a whole. An understanding of the parts may also be useful, but the emergent behavior could never be predicted from simply studying each part in isolation from all the others. It is yet another confirmation that things are interconnected in ways that we are only beginning to understand.

One example of emergent behavior is the movement of a flock of birds. Although the example of vengeful birds I described at the beginning of this chapter was based on fiction, it captured a behavior that can be observed on regular basis. Living in Michigan, it is common in the fall to observe massive flocks of birds in preparation for their flight to warmer southern climates. Although these birds are normally solitary or hang out in small groups, something triggers them to flock together at a particular time interval each year. Once they flock, the movement is incredibly complex and coordinated. Thousands of individual birds can flow together and change direction at the same time in a coordinated movement that defies explanation. Yet no single bird is in charge. The behavior simply emerges from the group. Various explanations had been considered and rejected when, in 1986, computer scientist Craig Reynolds programmed the basic rules of bird motion to create a simulation that closely resembled this emergent flocking behavior.[117] As it turns out, if each individual follows a few simple rules such as moving in the same direction as others, maintaining a minimum distance from the others, and avoiding predators, the complex flocking behavior will simply emerge. Yet no single bird is in charge.

The property of emergence is a key concept at all scales and systems, up to and including Reality itself. After all, Reality is the ultimate complex system. As a Buddha Scientist might observe, non-duality is the emergent quality of Reality itself. As such, *Buddha Science* would say that non-dual Reality can never be fully understood by measuring bits and pieces of it. Non-dual Reality is therefore indivisible because none of the parts, by themselves, can possibly have the emergent quality of non-duality.

In fact, it is clear that at some scales, the very act of measurement itself must be considered to be part of the system.[118] Once again, we can think about an infinity mirror. By simply looking into the mirror we are changing the image in it. It is certainly possible that this interdependency between observer and observed is true for measurement at all scales,[119] even though the effect of observation may be so infinitesimal at larger scales that it is not measurable using current techniques. In other words, if we look at a bird, or a waterfall, or the moon, is it possible that we are changing that observed scene in some miniscule and immeasurable way? This is consistent with the understanding, through *Buddha Science*, that the presence of a separate observer and observed is merely a concept, and has no meaning with respect to non-dual Reality.

On the other hand, we can never escape from Reality, despite the vast array of mind-altering chemicals that have been developed for the purpose. The best we can do is to try to understand it from our limited perspective. We can't simply step outside of Reality and measure it, so we settle for the blind men's approach of observing what we can and trying to see where our various views converge. This leads us again to the two approaches discussed in this book: the empirical observations and inference of *Buddha Science*; and empirical observation, conceptualization, and testing through the scientific method. So now let's begin to take a look at some additional areas of scientific inquiry, keeping in mind that the principles of chaos and complexity may apply to any and all of them.

CHAPTER SIX
PHYSICAL SCIENCE

Two theoretical physicists are lost in the mountains. The first physicist pulls out a map and studies it carefully, looking back and forth a number of times from their surroundings to the map, then turns to the second physicist and says: "I figured it out. I know where we are."
"Where?"
He points out toward the north. "See that mountain over there?"
"Yes."
"THAT'S where we are."

IT'S ALL RELATIVE

I went to college to shake a pesky habit of using common sense. In fact, one of the early lessons instructors often teach in basic science courses is that trying to apply common sense to scientific study introduces bias into the results. The scientific revolution brought about by the discoveries of chaotic systems, relativity, and quantum theory continues to challenge common sense, and has forced scientists to rely more and more on mathematics, models, and maps to understand Reality. In fact, if the two physicists in the above example exist at the quantum level, it is theoretically possible that the first one is correct: they are in two places at the same time.

Science and *Buddha Science* agree that what we think we know about the world—our common sense—is an illusion. While common sense tells us that we can swat a mosquito, eat a taco, or catch a fish without affecting anything else, we have already seen from chaos theory that every action has an effect. And *Buddha Science* tells us that it is also an illusion to hold the common sense notion that things we can see and touch are somehow permanent. Sometimes, we must let go of our long-held beliefs in order to gain a deeper understanding. According to Buddhist tradition, the Buddha used the analogy of using a boat to cross a river. When you get to the other side of the river, it makes no sense to try to carry the boat with you. The boat has served its useful purpose and it's time to let it go and move on. In the same way, we may not be able to continue a journey of understanding without moving beyond ideas and concepts that brought us this far.

We have already discussed chaos, and now it's time to tackle the other two recent and revolutionary discoveries of science: relativity theory and quantum mechanics. After discussing these two topics, we will attempt to take a look at how they relate to each other. Finally, we will show how the recent discovery of a new sub-atomic "particle" is raising even more questions in physics. These discussions challenge many concepts of how the world works, and new discoveries continue to raise fresh questions on an ongoing basis. We must approach these topics with a reminder that we are simply discussing ideas, maps, and fingers. They are only ways to try and gain a closer understanding of Reality. None of them is Reality itself.

Although Albert Einstein is credited with creating the theory of relativity, many scientists before him had built the conceptual and mathematical framework that he used to develop his theory. As a Buddha Scientist might say, everything is interconnected, even ideas. Such things as the invariant speed of light, size contraction with speed, and different clock speeds were known to science before Einstein published his famous theory. The mathematical concept of space-time itself was developed by Einstein's former college teacher, Hermann Minkowski.[120] In fact, early writings of *Buddha Science* describe the interpenetration of space and time.[121] The idea of spacetime is simply that space and time are interconnected, and so rather than referring to them as two separate things, he dropped the hyphen and smashed the terms together to create the word spacetime. Einstein was acknowledged in the scientific community for pulling many previously described concepts together into a "relatively" simple theory.

Relativity begins with the observation that the notion of motion of an object is meaningless except in relation to other objects in the universe. If nothing existed except two spacemen moving past each other, each spaceman would perceive his self to be still and the other one would seem to be moving. If there was also a fixed point in space, an observer watching the two spacemen from that point would perceive that both of them were moving.

The following passage was written by Nāgārjuna, a Buddhist philosopher, in the first century of the Common Era:

Just as a moving thing is not stationary,
A nonmoving thing is not stationary.
Apart from the moving and the nonmoving,
What third thing is stationary?[122]

Although Nāgārjuna's frame of reference was different from Einstein's, there is basic agreement here. Nāgārjuna's point is that motion is not an intrinsic property of a thing, but a relation between the thing and its position at different times, and this can only be measured by imposing an artificial system of measurement to its position. I don't know if Einstein was aware of the writings of Nāgārjuna, but we could speculate that the seeds of relativity began, not in the 20th century, but nearly 1900 years earlier.

Einstein divided the concept of relativity into two cases, that of special relativity and general relativity. He idealized both cases by ignoring complicating factors such as friction and hidden forces, and assigned a special "fudge factor" called the cosmological constant to help make everything fit mathematically. Of course, as we saw in our discussions of chaos, just because these complicating factors mess up our idealized version of the world doesn't mean they don't exist. Einstein himself called the addition of the cosmological constant to be his "greatest mistake," but scientists are still trying to understand what this constant may be telling us. One other thing that both the special and general cases of relativity share is that they both take place in four dimensions—three dimensions of space, and the fourth dimension of time.

Special relativity is limited to the case where things are in uniform motion relative to each other, and so they are in a perfect vacuum, with no forces or other masses acting to cause anything to accelerate or decelerate. This theory starts with nothing in the universe but a source of light and two observers moving past each other, and makes two basic assumptions. First, in this no-force world, the laws of physics are the same for all observers. As our two observers move past each other, because they feel no force of acceleration or gravity, each observer will feel as if he is still and the other one is moving. Second, the speed of light will be the same regardless of the relative motion of the two observers or the motion of the light source. These basic rules lead to a number of counterintuitive conclusions. Here's where you leave common sense at the door. Don't lose track of where you left it, because you may need it later.

Let's use an example of an intergalactic superstar basketball player who we will call our Star Player. He's standing in the middle of a magic basketball court that is moving to the left with respect to the stands. Although the court moves extremely fast, it is otherwise pretty standard, with a game clock and baskets at each end. Our Star Player is standing center court, dribbling a ball of light. From the Player's perspective, the light ball is bouncing vertically up and down, and moving, of course, at the speed of light. But from the viewpoint of Superfan sitting on the stands watching the court go by, the ball of light will not only be moving up and down, it will be moving forward in the direction the court is moving. So the path that the ball travels will appear to Superfan to be v-shaped. The path that Superfan sees will therefore be much longer than the vertical up and down path that Star Player sees. This difference in the apparent motion of the ball, though based on a train analogy described by Albert Einstein, echoes the observations of Nagarjuna, who concluded that motion, mover, and route are non-existent,[123] and defined only by the relationship between them.

Using common sense, since it takes the same amount of time for the ball to go up and down no matter where you are watching from, and Superfan sees it travel a longer path, it would appear to Superfan that the ball is moving faster than speed that Star Player sees. But remembering the second rule of special relativity, the speed of this light ball has to be the same from either point of view. So what gives? How can the ball travel a longer path at the same speed and still arrive in the same length of time? In order to resolve that question, special relativity says that from Superfan's perspective, the game clock is moving slower than the time he shows on his smart phone. So instead of the speed of light increasing, the "speed of time" has decreased! Although the Player started dribbling at 1:00 and the clocks matched then, now Superfan's phone is showing 1:05, and the game clock is showing 1:04. This is the first confusing result of the special theory, that moving clocks appear to run more slowly than clocks that are not moving. Another way to think about this is that everything moves through space-time at the same speed, so if it moves faster in space, it moves slower in time.[124] This result is known as time dilation, because it is not just about clocks, it is about time itself. And Einstein was right. Experiments have confirmed that time moves more slowly for things moving in space with respect to an observer.

Next, our Star Player takes two balls of light, one in each hand, and throws them toward the baskets on either side of the court. In an incredible demonstration of prowess, both balls sink like butter into their respective baskets at exactly the same time, at least from the viewpoint of the Player. But Superfan sees something different. Both balls fly at exactly the same speed, but since the court is moving to the left, as the ball flying to the left approaches the basket, the basket is moving away from it so it takes a long time for this ball to sink into the net (nothing but air). The ball on the right, however, finds the basket almost immediately, since the basket is moving toward it. So while our Star Player sees both balls sink simultaneously, Superfan sees the right ball sink first and the left ball sink much later. This is known as the principle of simultaneity, which says that the idea of simultaneous events is relative to the observer. Sounds crazy, doesn't it? But it's true.

I told you to check common sense at the door. The problem is that we take the common-sense view that time is something linear that moves ahead no matter where we are or where we are going, but this view is only an illusion. Once again, our concepts get in the way of experiencing Reality as it is. Einstein himself said that time is not a condition in which we live, but only a way that we think about it.[125] *Buddha Science* would agree.

Another result that follows from special relativity is called length contraction. From the viewpoint of our Star Player, the court measures the standard NBA length of 94 feet. But to Superfan, the court looks much shorter. In fact, this is consistent with the measurements that the Player is making, because from the viewpoint of Superfan, the Player's measuring stick has shrunk by the same ratio as the court. Don't laugh—I meant his ruler. In fact, the Player looks much slimmer to Superfan as well. So forget all of the ads for special diets—the simplest way to look slimmer is to move really fast. And it's instantaneous—no need for a gym membership. But it's not just the player and court that are compressed; it's also the space around them. Once again, not only is the idea of linear time an illusion, but the fixed dimensions of space are as well. Like the blind men and the elephant, both the Player and Superfan are correct, even though they both experience different things. It all depends on their framework.

What about matter itself? Isn't the mass of something real, solid, and invariant? Sorry, but Special Relativity does away with that notion as well. In the most famous equation that is related to Special Relativity, Einstein showed that the mass of an object or system is directly related to its energy content. The faster Star Player moves, the more massive he becomes. Another way to think of this is that mass does not equate to substance; it is simply a form of energy.[126] This finding is absolutely in accordance with the *Buddha Science* observation that everything is process, and that solid and unchanging "things" are merely illusion.

So why do things seem solid to us? Why do we feel the bump when we hit our elbow on the table or hear that nice "thwack" when we hit the golf ball? Actually, in my case, the golf club misses the ball completely more often than not, but if you get down to atomic level, even when I hit a good shot there's not a solid club hitting a solid ball. Instead, the electrons and protons in the club repel the electrons and protons in the ball, causing a series of similar actions within the ball itself, in turn causing the ball to be repelled down the fairway.

Nothing at the atomic level has actually touched anything else; in fact, there's no solid matter involved, since the electrons and protons themselves are merely clouds of energy.

The equation that Einstein developed states that the energy in a system is equal to its mass times the speed of light squared ($E=mC^2$). Since the speed of light is 186,000 miles per second, this calculation results in a huge value for the energy in a system. One result is that splitting an atom with very tiny mass results in release of the incredible energy we know from atomic bombs. The key observation is that mass and energy are two different forms of the same thing. And the energy in the system can take a number of different forms.

So the changing mass referred to as a result of special relativity comes from kinetic energy, which is the energy of motion. Anything that is in motion has more kinetic energy than the same thing at rest. The Einstein equation shows that adding energy of any kind to a system increases the mass of the system, so setting an object in motion with respect to an observer also increases its mass. The faster our magic basketball court is moving, the more massive our Star Player will appear to Superfan (even though he also looks thinner). Another way to think about this is to observe that it takes energy to accelerate a mass to a higher speed. The closer a mass comes to the speed of light, the more energy it takes to accelerate it to the speed it is traveling. In fact, it takes an infinite amount of energy to accelerate a mass all the way up to the speed of light, which makes it impossible. This is one reason Einstein said that the speed of light is the maximum speed that anything can travel. Light travels at that speed because it has no mass.

To recap, from the two simple assumptions of special relativity, space, time, and mass are all relative, based on the framework that you view them from. Each of these properties is intimately related, or as a Buddhist Scientist would say, interconnected, with all of the other properties. If you specify any of these properties, it is meaningless without putting it into the context of the other properties. The result from special relativity is that when the intergalactic basketball court moves past the stands, Superfan sees time moving slower on the court, a thinner yet more massive Star Player, and a different sequence of events than Star Player will see.

Now let's take a look at the second part of this theory, general relativity. As we have said, special relativity deals with the case of uniform motion—that is, with Star Player's magic court moving in a straight line at exactly the same speed for the whole demonstration. General relativity adds acceleration to the picture. So now let's imagine that our Star Player is launching his universe-wide victory tour, starting in his hometown of Basketball City, and Superfan is on the platform, seeing his hero off. The magic court starts from the platform and accelerates up to nearly the speed of light. He does a "fly-by" tour of the universe, and then returns to the end of the tour back at Basketball City.

General relativity starts with the understanding that all of the principles and results of special relativity still hold, and adds two principles that are due to acceleration. The first, called the principle of equivalence, states that accelerated motion and being at rest in a gravitational field are physically identical. In other words, as the magic court starts to move to the left, our Star Player experiences a force exactly like a gravitational field trying to pull him to the right. Star player throws the ball up into the air just as the court accelerates to the left. From his point of view, the ball falls off to the right and is lost forever. From Superfan's point of view, the court would move out from underneath the ball. It's a good thing that Star Player has a lot of balls, or the tour would be ruined. And the same thing happens after the court returns from the big tour. As the court pulls into the station from the left and puts on the brakes, the deceleration threatens to make everything to fly off to the right. In both cases, the court acts exactly as if there is a gravitation force from the right, trying to pull everything toward it.

Let's stop for a minute to try and understand, intuitively, why this happens. In the first case the court is accelerating. When Star Player throws the ball up, the court is moving to the left at, say, 1,000 miles per hour. From the point of view of Superfan, before the ball comes back down, the court has accelerated to 2,000 miles per hour, but the ball is still only moving at 1,000 miles per hour, so the court has moved out from under the ball. From the point of view of Star Player, it looks like the ball arcs to the right in the air, is if caught in a gravity field. Coming back from the tour and moving to the right from Superfan's perspective, when Star Player throws the ball up in the air, the court is moving at 2,000 miles per hour. The ball continues to move at this speed, while the court decelerates to 1,000 miles per hour, so the ball flies past the court, once again off to the right. From the point of view of Star Player, court acceleration as it moves left and deceleration as it moves right produce exactly the same result. It looks to Star Player that the ball is caught in a gravity field that makes it arc to the right, exactly the same as what he saw as he left on the tour. From Superfan's point of view, the two actions are directly opposite. So who is right? Like the blind men, both of them are, based on their particular perspective of events.

The second principle of general relativity, which we call spacetime curvature, is that gravitational force is not due to a massive body (like a planet) with some inherent property called gravity; it is because a system that has mass and energy causes a curvature in spacetime, and this is what we experience as gravity. So despite what you learned in grade school, the earth doesn't suck. This spacetime curvature forms what is known as a gravity well around the massive object. The more massive and energetic the system is, the deeper the gravity well is and the harder it is to escape from it. Also, the closer you are to the massive system, the stronger is the force trying to pull you in. You have probably seen the funnel-like structures at some science fairs or museums where you roll a coin around and it spins faster and faster until it goes down the hole in the center. Let's say you accidently tossed your lucky coin into one of these and so you tried to grab it back. You might have some chance of getting it on the first two circuits, but as it gets closer to the center and moves ever faster, it would be harder and harder to retrieve the coin.

Two things: first, I'm not advocating that you try to grab coins out of these funnel-things, especially if they aren't your coins. Second, check before you toss a coin in to make sure it's not your lucky one. The point is that as you approach the massive object, or as acceleration increases, so does the force that you, or the coin, or Star Player, feel. If our magic court hits the brakes too hard on the return trip, Star Player will fly off the court and be lost forever. Imagine the victory tour promoter trying to explain that one to his insurance agent.

So general relativity adds gravity-acceleration equivalence and spacetime curvature to the mix. Some very strange effects result from adding these two principles to what we already know from special relativity. We already know that time does some pretty strange things in the world of relativity. General relativity says that the more gravitational force you experience, the slower your clocks will run. This principle is known as gravitational time dilation, and the effect is real. For instance, the clocks in GPS satellites run faster than clocks in the International Space Station. Why is this? Because GPS satellites are in a higher orbit than the International Space Station, so the GPS clocks experience less gravitational pull than the clocks on the International Space Station. This is equivalent to less acceleration, so it translates to faster time and the GPS clocks run faster. In fact, this effect holds at all scales. So when you're standing or sitting up—anytime your head is higher than your feet—your head is actually moving faster in time than your feet are.[127] They say tall people generally get the best jobs and opportunities. Perhaps that's because they are always "ahead" of the rest of us. If it makes me think faster, from now on I'll climb up on the roof to do Sudoku.

Getting back to our Star Player, when he does his intergalactic victory tour starting from Basketball City, he has been exposed to acceleration and Superfan has not. As we have said, less acceleration means faster time. So when he gets back to Basketball City, he will find that even though the tour lasted two years according to the court clock, the clocks in Basketball City have moved much faster and Superfan has long since died of old age. Star Player will be greeted by Superfan's descendants.

We won't elaborate on the other effects here, but they include some very strange things. For instance, the curvature of space-time means that rays of light bend around very massive objects—an effect confirmed by observation through powerful telescopes. Another effect is that the universe is expanding, and although not everyone agrees, this expansion is now thought to be accelerating, possibly due to something called "dark energy." We will be discussing this later on. More complicated effects involve changes in the "wobble" of planetary orbits, and something called "frame dragging," where space-time itself becomes twisted by rotating masses.

The thing is, no matter how we conceive of it, the world of relativity is the most widely accepted current concept of the actual physical world in which we live. This concept may change in the future, but it's how science currently understands physical reality. The phenomena predicted by Einstein have been shown to exist. All of this may seem to contradict common sense, but the problem is not what the world does, it is the concepts we form around it. The universe is the way it is, whether we like it or not. My basic scientific training was correct—let go of common sense. The Buddha would approve. Common sense is only the boat that allows us to deal with the day-to-day world. To gain a deeper understanding of the world, we must leave that boat at the shore.

BUDDHA MEET SHRODINGER'S CAT IN A BAR

As we have seen, Einstein left the shore of common sense to investigate a world where things move very fast. Now we turn our attention to a world where things are not only fast, but very, very small: the world of quantum physics. In fact, things at this level are so small and fast that scientists don't actually see them directly; they see the tracks left by them. By the time they are aware of an event in this tiny world, the event has already passed. Physicists study images on a plate or tracks in a cloud chamber to interpret the behavior of things that occurred split seconds before. In this sense, quantum physicists study *schul*, a Tibetan word for the impression of something that used to be there.[128]

At the broadest level, the world of quantum physics shows us two very distinct levels of reality—that of the day-to-day observations where things seem definable and solid, and a reality of undefined potential, where individual things do not exist, and even the ideas of space and time are only illusions. Once again, this is a realm that we can only try to understand by forming concepts, but those concepts simply allow us to glimpse bits of this Reality; they can never fully capture it. This quantum Reality underlies and is included in everything we know, measure, understand, and experience. This finding of modern science is amazingly similar to the two-world observations of *Buddha Science* that were made centuries ago.

Scientists themselves admit that they don't understand the world of the quantum. Perhaps the most famous quote along these lines, generally attributed to Nobel laureate Richard Feynmen is, "If you think you understand quantum mechanics, you don't understand quantum mechanics." Unlike Relativity, the quantum world is not fully described by a well-articulated theory. The basis for modern scientific ideas about the quantum world is mathematically defined in four papers written in 1926 by Erwin Schrödinger.[129] These equations assume that specific conditions will yield specific, well-defined results, but combining them with experimental observations shows an intrinsic randomness.[130] Since experimental results do not match those predicted by the equations, scientists use different ways to interpret the results to try and account for the mismatch. There have been many attempts to interpret experimental results to try and make them fit the mathematical model defined by these equations. Earlier on, we talked about the tendency of science to try and use existing models until they just don't make sense any more. Although someone may find an interpretation that fits without causing all sorts of other problems, it's certainly possible that eventually, the current model may have to be modified or abandoned.

There are literally dozens of interpretations, some reasonably plausible and others that sound pretty much like science fiction. Some begin from the assumption that the quantum world would be logical and predictable if we are able to discover new rules that will clear up some questions. Others insist that there is a deterministic world "down there," but our observations disturb the system so much that we can never know exactly what it looks like. Yet others propose that the world at this level is literally undefined until we observe it, so the very act of conscious observation causes the non-dual world of undefined potential to collapse into the dual world of our everyday experience. To paraphrase yet another former president, Ronald Reagan, "Here we go again." Once again, this sounds a lot like the two different realities of *Buddha Science*. In fact, some researchers even think that science should abandon any attempt to explain the problem, relegate it to metaphysics, and just accept that it works.[131] After all, many of the interpretations can't be proven false, so they don't meet the basic test of science laid out by Karl Popper.

The basic problem of the quantum world came to light in 1875 when Max Planck determined that electrons could only emit energy in discrete bursts, called quanta. Until that time, the atom was considered to be just like a planetary system, where a planet could form any orbit around its sun. Think about riding a bike in circles. You can make the circles any size you want, depending on how much you turn the wheel, how fast you pedal, and how much you lean. But if you were a bicycle-shaped electron circling a huge nucleus, you might only be able to ride in circles with a radius of, say 10 feet, 14 feet, and 23 feet. Every once in a while you would be instantaneously transported to another one of the three "allowed" circles. Your riding behavior would be "quantized."

Planck initially suggested that this effect could explain why very hot materials glow different colors as they heat up. As energy is added to the materials, the electrons jump to higher levels and the color changes. Then Albert Einstein suggested that light also was quantized into discrete elements he called photons. This was rejected by the scientific community at the time because light was known to be a wave.[132] Waves are continuous, and the idea of discrete quanta screams "particle!" This set up the debate that still continues today. Although science is showing us that our ideas don't quite fit, we still can't shake the idea that "energy" and "stuff" are two different things. Light is energy, and energy comes in continuous waves, right? And particles are "stuff," which means that there is one here, one there, and they are two separate things. Those are the only two choices, right? From the viewpoint of *Buddha Science*, this is our conceptual world getting in the way of what we actually see. Things are much more interconnected and complex than this simple view of the world suggests.

In a sense, the scientific community acts a lot like a quantized electron, jumping back and forth between wave or particle ideas, unable to find an orbit in between the two. Once again, the map is not the reality. Just because waves and particles are the only two things we can conceive of to explain the reality we see, that doesn't mean that reality is obliged to conform to our conceptions. Still, it's worthwhile to describe the concepts that experimenters try to test in order to understand why things are so confusing in the quantum world. First, let's discuss the concept of waves.

Go out to a pond on a calm day, chuck a couple of pebbles into it, and watch the patterns they make. Note that I'm not suggesting you use your neighbor's decorative fish pond for this purpose. At any rate, you will see a wave circle radiating out from the point where each rock went in. Where these wave circles intersect, you will see an alternating pattern of peaks and troughs where both waves meet. In some places, the two waves will combine to form a high peak, and in others, the two will cancel to create spots of smooth water. This is called wave interference, and it results from both waves existing in the same space at the same time. This ability for two things to occupy the same space and time, called superposition, is considered to be a property of waves. This ability is also recorded in many experiments with what seem to be subatomic particles, which is the basis for much speculation. Are these particles that show wave-like properties, waves that collapse into particles, or something else entirely? If they are particles, how can two particles occupy the same place at the same time? If they are waves, why do they look like particles when we measure them?

The term *superposition* refers to more than one thing occupying the same place at the same time. The idea of superposition sounds very scientific and exotic, but in fact it is something we experience every day with regard to light. Superposition is the reason we have rainbows. What we experience as white light is simply different wavelengths, or colors, of light that all exist at the same place at the same time. Water droplets in the atmosphere scatter this light into the different colors we experience as a rainbow. It this way, the rainbow shows us the different colors of light that are in the same place at the same time in the atmosphere. The thing that is so startling is that experiments show that matter can act the same way at the quantum level. So at the quantum level, different bits of matter can occupy the same place at the same time. As Einstein showed, matter is equivalent to energy, so why can't matter act like a wave, while energy can act like particles? The confusion we experience is based on our concepts of waves and particles and what each can and cannot do. A physicist observes an interference pattern (like the one from overlapping waves in a pond) and applies the belief that this pattern represents a wave phenomenon, because that is what he is familiar with. Since he cannot conceive of another alternative, he assumes that no other alternative exists. All we really know is that we measure interference patterns and observe superposition. The explanations that we use to explain these observations are conceptual, and as we know, concepts are not Reality.

Superposition is something that exists, in spite of the fact that we cannot fully conceive of how it works. Quantum computers are being developed that take advantage of superposition. Instead of a "bit" being either 1 or 0, so-called "cubits" can be 1 or 0, but they can also be both 1 and 0 at the same time. This may not seem like it would make much difference to computing speed, but by stringing cubits together, quantum computers could potentially perform certain types of calculations thousands of times faster than conventional computers.

Erwin Schrödinger was an Australian physicist who developed a number of mathematical equations critical to quantum physics. He is most famous, at least among the general public, by creating a thought experiment to describe superposition. A cat is placed in a box with some equipment that kills the cat if a quantum particle enters the box and trips the mechanism. In the superposition state before you open the box to see what has happened, the cat is both alive or dead, and/or neither, depending on your point of view. Once you look into the box, the cat can only be alive or dead. The act of observation has collapsed non-dual Reality into the dual situation of a cat that is either alive or dead, but not both. The disclaimer says that no cats, quantum or otherwise, were killed or brought to life during this thought experiment.

This thought experiment is actually a pretty good description of the Buddhist view of two levels of Reality. Non-dual Reality, a state of unlimited and form-less potential, is collapsed by our concepts into the dual world that we experience. Although I said our discussion would be limited to observations and direct inference based on observation, I can't resist the temptation to slip into the realm of metaphysics, just this once. In early texts describing the death of the Buddha's body, he is said to have passed into Nirvana, where he was released from the endless cycles of life and death. In the terms of quantum physics, the Buddha actually reversed the process of observation, moving from the dual reality of our daily world into the formless potential inside the unopened box. Because he had let go of the dualistic idea of self, he became a non-self, neither observer or observed. In other words, the Buddha became Schrödinger's Cat! In fact, anyone who is fully enlightened becomes such a cat when their body dies. I hope it's a big box, because it's getting more crowded all the time. On the other hand, since things in the box are in superposition, I suppose the box doesn't need to be very big at all, because all of the new Buddhas could exist in the same space.

Another behavior that has been recorded in the quantum world is called *entanglement*. This is not a question of the behavior being conceptually attrib-uted to something familiar like wave or particle behavior, but something that seems completely impossible in our ordinary world. Entanglement basically means that pairs of particles can act as a system, so changes in one of these particles also means that the other particle is changed. Could this be a matter of emergent behavior in the way that a school of fish behaves, as we discussed in the last chapter? Although it is certainly possible that entanglement is somehow related to the system as a whole, it is distinctly fishier than that, because it seems to actually violate the basic laws of physics as we understand them. In particu-lar, it seems to violate the speed of light.

In a way, entanglement is almost a corollary of superposition—instead of two particles sharing the same space, it could be thought of as two particles sharing the same existence while being separated in space.[133] Note that I didn't say the particles are related, but that they seem to share the same existence. Why would physics describe it that way? The spooky thing about this behavior is that it can happen instantaneously, regardless of the distance between the particles.

Somehow, it seems, entangled particles can communicate faster than the speed of light. Since conventional physics tells us that's impossible, there are limited ways to explain how it can be true. Once again, this is not simply a theoretical phenomenon. Researchers in China have successfully teleported entangled photons over a distance of 88 miles. When the sender measures the quantum state of their photon, the receiver's entangled photon changes instantaneously.[134] It takes longer for the humans to find this out, because the communication between the two sites is limited by the speed of light. Researchers at the site that sent the entangled particle immediately communicate that to the receiving site. But by the time the receiver site gets the word that a change has occurred, the particle on their end has already changed. Since a single entangled particle can exist at more than one place, quantum physics shows us yet another, more intimate, level of interconnection than we have seen before. Although these experiments are only a glimpse at the future, it is possible that this is the first step toward realizing teleportation of the sort envisioned in the Star Trek television series. Hopefully, during the experimental phase of this technology, we will limit our creation of human/fly hybrids to a minimum.

So entanglement can be viewed either as instantaneous communication through space, or as a single system that exists in two places at once, like our physicists at the beginning of this chapter. In the previous chapter, we spoke about the concept of emergent behavior—the observation that complex systems can show behaviors that can never be understood by simply looking at the parts. Entanglement suggests that there is a larger, more complex system at work, a larger Reality. Our current concepts of space, time, and movement simply cannot explain entanglement. As *Buddha Science* would say, things are interconnected in ways we may never understand. Is movement real, or is our concept of space and separateness merely an illusion? From the perspective of non-dual Reality, the following quotation from the Flower Adornment Sutra seems particularly applicable: "Going and returning with no border, movement and stillness have one source."[135] The source this Sutra refers to is the Tao, what we are calling non-dual Reality.

As *Buddha Science* tries to point the way toward non-dual Reality, science tries to discover how it works through combining relativity and quantum mechanics into a unified field theory. As science once split into different disciplines to study smaller and smaller parts, there is significant current movement toward combining the parts to understand the dynamics of larger and larger systems. If all of the various scientific studies are envisioned as a tree, unified field theory can be viewed as the trunk, the ultimate combination of universal laws.[136]

A key to that combination will be a better understanding of time itself. As we have seen, relativity relegates time to one dimension, mathematically equivalent to a single direction in space. Scientists talk about the arrow of time, because it points in only one direction. There is no mathematical reason that time needs to only move forward, but probability theory strongly points to a tendency for it to do just that. Where relativity theory begins to have problems is the assumption that time moves forward in a linear manner, which leads to the limit on the speed of light. Quantum physics is also bedeviled by time. If our experiments show us what we design them to see, and entanglement can cause simultaneously effects in spatially separate particles, perhaps we need to rethink our concept of time. What if time itself is not inflexible and linear? What if it is a not a reliable way to measure velocity? What if time is chaotic?

Johns Briggs and David Peat have suggested that time may be a strange attractor.[137] Psychologically, we experience some moments that seem to fly by, and others that move with agonizing slowness. Are we actually experiencing time as it is in those moments, free from the concepts that our clocks impose on us? Since strange attractors repeat patterns, could this explain why history tends to repeat itself? Time is one of those scientific and common sense concepts that may need to be revisited in order to tie together the observations of relativity and quantum mechanics.

But why does time always move forward? After all, the underlying laws of physics work the same going forward or backward in time. In 2009, researchers at the University of Bristol in the United Kingdom reported that they had tied the forward movement in time to quantum particle entanglement, suggesting that even such familiar things as a cooling cup of tea happens because it becomes increasingly entangled with the air around it.[138] As the universe moves forward in time, eventually, everything becomes entangled with everything else. Substitute the word "interconnected" for "entangled," and this sounds exactly like *Buddha Science*. Some logicians believe that the only difference between the past and the future is that we know more about the past than we do the future. Other than that, past and future are equally real. Some forms of *Buddha Science*, on the other hand, suggests that nothing is real except the present, and the past and future are merely illusion—a human construction that we use to try and make sense of the present. Do we use the arrow of time to shoot ourselves in the foot?

If everything eventually becomes entangled with everything else, how can we separate anything from anything else? One observation that relativity and quantum mechanics have in common is that any observed phenomenon cannot be separated from the observer. This confusion of observer/observed is captured in a quote generally attributed to Neils Bohr: "A physicist is just an atom's way of looking at itself." Just because our minds may tell us that the moon is still there even if no one is observing or measuring it, we cannot be certain that this is true. As quantum mechanics seems to indicate, it is certainly possible that the act of observation itself is what actually brings particles into existence. This, of course, matches the *Buddha Science* observation of two realities, including the "inside" perspective of duality, which indicates that there can be no observed without an observer.

There are two significant recent discoveries that may help scientist in their quest to synthesize these concepts. One is the discovery of the Higgs boson. The other is the discovery of evidence for gravity waves associated with the big bang. Together, these may help scientists to understand better how mass and energy are related.

The Higgs boson provides evidence for an idea developed over 40 years ago by a team of scientists led by Peter Higgs. This idea envisions a uniform energy field called the Higgs field, which exists everywhere throughout the universe. As the boson travels through this field, the disturbance in this field imparts mass to all of the particles around it.[139] The discovery of the boson itself will help scientists understand the actual mechanism for converting energy to mass according to Einstein's equation. In turn, because photons have no mass and particles of matter do have mass, it will provide a link between matter and light itself. Are matter and light merely different aspects of the same thing? Certainly, this would be consistent with non-dual Reality.

The second recent discovery, that of gravity waves, was based on images from a telescope known as BICEP2. This telescope measures something called the cosmic microwave background, which is remnant energy from the big bang, the event that formed the universe as we know it nearly 14 billion years ago. Earlier in this chapter, we introduced the term *schul*, which means the impression of something that used to be there. This telescope records patterns in the cosmic microwave background that scientists have interpreted as *schul*, or "footprints in the sand" left by gravity waves that occurred as a result of the big bang. By looking closely at these footprints, physicists can try to deduce how fast, and how energetically, expansion of the universe occurred, which will help them to limit the number of viable models of this expansion.[140] This should also help physicists to better understand the properties of the subatomic particles that were created as a result of the big bang, and eliminate some interpretations of quantum mechanics that we discussed earlier.

We should note here that *Buddha Science* rejects the notion that the Big Bang was the beginning of anything. It asks the question, "How could something be created from nothing?" *Buddha Science* says there is no beginning or end, but endless cycles of existence. So in this view, although the big bang may have occurred, it was not the beginning of everything but merely the beginning of another cycle. In fact, today's physics cannot determine what the state of things was at the moment of the big bang because of something called the "Planck wall." This wall limits our ability to understand what happened in the first 10^{-43} seconds after the big bang. Before that time, gravity did not exist as we know it today, so our current physics has no tools to deal with it.[141] The position of *Buddha Science* is therefore compatible with our current science, even if points to the fact that science doesn't know the answer. I'll have to keep that argument in mind the next time my wife accuses me of not listening. I'll just say, "Sorry, but the Planck wall doesn't give me any way to analyze that statement." That'll work, right?

Although the maps we have created of the universe do not yet look like the terrain, through science we now know that an event occurred that caused rapid expansion of the universe, and that as it expands, the universe is becoming smoother and more uniform. Rather than an explosion that would cause disorganized scattering of materials, it was more like a birth, echoing the following description from the Tao te Ching:

> *There is a beginning which contains everything. It pervades everywhere unhindered. It may therefore be called the world's mother.*[142]

These new discoveries of science, in turn, may lead to a better understanding of something called dark matter and something even more mysterious called dark energy. Together, dark matter and dark energy make up 96% of the known universe, yet scientists don't really understand exactly what they are. Dark matter seems to be concentrated within galactic clusters, while dark energy is uniformly distributed throughout "empty space." This concept of dark energy is consistent with the *Buddha Science* observation that emptiness is not the same thing as non-existence. Something exists everywhere in space, even though we cannot say exactly what that something is. Although our understanding may improve, for now the majority of the universe meets the properties of a non-dual universe that cannot be fully described or understood.

The late David Bohm, a theoretical physicist and philosopher, proposed the concept of a universal flux that cannot be explicitly defined or fully known, in which mind and matter are not separate substances, but different aspects of one whole and unbroken movement.[143] He spoke of our concepts to try and define this universal flux as being limited by the subject-object structure of language.[144] Although Bohm argued against adopting the eastern approach of meditative observation of wholeness,[145] his descriptions of "wholeness" sounds strikingly similar to the following description of the Tao, or non-dual Reality:

The tao that can be told
is not the eternal Tao
The name that can be named
is not the eternal Name.

The unnamable is eternally real.
Naming is the origin
of all particular things.[146]

Even though the two fingers, science and *Buddha Science*, may be pointing toward a common view of what our universe is and the basic building blocks that underlie it, we need not go soaring into the universe or shrink down to subatomic size to find mystery. One of the most profound mysteries awaits: that of life itself. Trying to sort out the distinction between energy and mass, time and space is one thing, but how do we understand life? What is life, exactly? How did it come about, and how did it become what it is today? In the next chapter, we will begin pointing our fingers toward the mystery we call life.

CHAPTER SEVEN
SCIENCE OF LIFE

One of the most poetic facts I know about the universe is that essentially
every atom in your body was once inside a star that exploded.
—Lawrence Krauss, in *A Universe From Nothing*

THAT'S LIFE, I CAN'T DENY IT

Let's begin with the most fundamental question: What is life, exactly? The question is more complex than it appears at first glance, because next we have to ask if we are trying to define life as we know it exists on earth, or life that potentially could exist in other forms. The question is critical in the search for life outside of our solar system. In the language of physics, the term *life* is meaningless except in relation to descriptions of biological phenomena.[147] Peter Ward, a professor of biology, earth and space science, and astronomy at the University of Washington, has explored the question in depth in a book entitled *Life as We Do Not Know It*. He takes the position that life can emerge from a mishmash of things such as cells, viruses, and genes that are not alive in and of themselves.[148] In this view, life is an emergent quality of a complex system, a concept we discussed in Chapter 5.

There's not a single definition of life that is universally agreed upon, although NASA generally accepts the late Carl Sagan's definition of life as a chemical system capable of Darwinian evolution.[149] The Oxford Dictionary (American English) defines life in a more limited framework as "The condition that distinguishes animals and plants from inorganic matter, including the capacity for growth, reproduction, and functional activity preceding death."[150] Other definitions use phrases such as "not including dead organic material," or add "the capacity for continual change" as a characteristic of living beings.[151] These phrases point out the difficulty of defining life as an abstract concept. At the level of plants and animals, generally "we know life when we see it," but when it comes to defining the boundaries of what actually constitutes "life," we run into all sorts of difficulty. When you come down to it, you can summarize most of the definitions of life as a fancy way to say "stuff that isn't dead."

After all, scientifically, there is no universally accepted reason to reject the idea that life needs to be based on organic material. What about crystalline forms of life, life forms based on dark matter, or life that is not matter at all, but composed purely of energy forms? Although such speculation sounds like science fiction, these are questions that are examined by the scientific community in the search for extraterrestrial life. If our view of life is too narrow, we risk failing to recognize life even if we do encounter it out there.

From the perspective of *Buddha Science*, life is most like the position of Peter Ward—an emergent quality that imparts being. The result of life is that the being must subsist on nutriment.[152] Without this nutriment, biological life will cease. Life as we know it on earth includes plants, animals, and insects, but beyond this, to my knowledge Buddhism does not attempt to enumerate which particular systems may be considered to be alive. In the Buddhist circle of life, one category is animals, which includes beings who respond only to pleasure and pain. I suppose that could be considered *Buddha Science*'s minimum criteria for something to be considered alive, although I'm not sure how you would determine if a common cold virus was feeling pleasure or pain. Actually, I'm not sure I want to know. At any rate, the division between life and non-life is something of a false division for *Buddha Science*, because life is considered to be merely one stage in endlessly turning cycles of life and death. There is no beginning or end to life, or to anything at all for that matter, only the endlessly turning cycles. Perhaps these cycles are one reason we have such a hard time defining what life is.

In the English language, the word *life* can mean either the general sense of life as a universal quality of certain things, or as the particular state of an individual organism. So, for instance, we talk about fossils as representing ancient life, although many fossils consist of materials that were never "alive" as we define an individual life. These fossils include shell or skeletal material, footprints, and the like. We refer to organisms and people who had the quality of being living moments before, as no longer having life. This dual usage of the word also leads to significant confusion and conflict when, for instance, we debate at what point a human fetus, which clearly has the quality of being alive in the environment that contains them, actually becomes an individual life.

Life in either sense requires balance, a chain of causes and conditions that allow life to emerge and sustain itself. We have a small pond at our house, about one third of an acre in size, and it is incredible to watch it because it changes constantly. I have been trying to avoid or minimize chemical treatments in favor of more natural solutions, but occasionally things get out of hand in terms of my particular human aesthetic. One year, the pond developed a duckweed problem to the extent that it looked like a golf green. I broke down and used a chemical that worked so well, I used it again the following year as a preventative, even though the duckweed problem was gone. It pretty much killed every bit of vegetation that touched the water, although the fish seemed happy enough. Well, that winter was brutal, and in the spring as the ice melted, virtually every fish in the pond came floating to the surface. Even though we knew a lot of people who had major fish kills in their pond that year because of the cold weather, I am convinced that some fish deaths in my pond were due in part to my careless change in its conditions for life. I decided to let the native weeds grow back up and do minimal spot treatments for algae, and about mid-summer the next year we started getting schools of minnow. By August, the pond was once again teaming with life. Walking along the edge would cause frogs to plop and the water to boil with fishy activity. Although large bass have died off, there is now an entirely new generation of creatures ready to continue the cycle. It is as if the pond itself is infused with the essence of life, which it imparts to the individual creatures within it. Despite my ham-handed efforts at treatment, the pond refused to die.

My view of the pond is essentially the concept of Gaia writ small. Developed in the 1960s by British scientist James Lovecock, Gaia theory treats planet earth as a living, interconnected system. He understood emergence before emergence was cool. In this theory, life itself is an emergent quality of this complex and dynamic system. The system itself is alive, with the individual organisms within it representing small patterns of life within that system. This system developed life long before it had humans around to act like they owned it.

Once life emerged on earth, it was just as tenacious as the life in my pond. This thing we call life continues to change, evolve, and grow, even though most individual organisms we recognize as alive have a lifespan of less than one year.[153] The lifespan of individual organisms does not limit the length of time that a particular species will remain viable. Consider the lowly but troublesome cockroach. The earliest fossils suggest these creatures have been around for 350 million years, yet their individual lifespans range from only around 30 days to two years. They have survived through the eons by rapid reproduction and evolution, and by the ability to adapt their behavior quickly to changing environments.

As a system, life produces itself. Reproduction is one of the primary characteristics of life as a way to perpetuate itself, regardless of the lifespan of any individual organisms. Paradoxically, individuals only exist in relation to the system itself—the individual is an illusion.[154] Individual organisms are merely patterns that exist within the ever-changing flow of life.[155]

But what are the limits of an individual life? If it's merely a pattern, when does that pattern begin and when does it end? The sprout is potential in the seed, so they are two stages of the same entity. In this sense, the sprout is self-caused.[156] But the sprout has the potential to mature and create more seeds, so the pattern of an individual cannot be separated from the broader pattern of flowing life itself. If you plant a vegetable garden, you know there is a time in the season when the plants begin to wither at the same time the vegetables are getting stronger. The plant is sacrificing its energy to create the seed that will continue the pattern of life. It is simply one part of the cycle that goes from seed to germination to plant growth to growth of the reproductive mechanism and then to seed again.

Patterns cross not only individual lives, but different life forms as well. Species-specific patterns contain the evolutionary history that resulted in that organism. Our bodies form three different types of kidneys in succession while we are still in the womb, each one similar to those found in different types of fish. Finally, these are replaced by the adult kidney before we are born.[157] Even in the womb, we follow patterns of evolutionary development from the ever-changing and interconnected flow of life.

Individual organisms are constantly changing patterns. We have already mentioned that all of the cells within a human body are replaced over a seven-year period, the chemistry of the pancreas changes every day, and the cells lining the intestines are replaced over a three-day period.[158] The body you have in this moment is, literally, not the same one as when you started reading this chapter. Moment by moment, our very life is dependent on substances from our surroundings. We cannot survive for long without the air we breathe, the water we drink, or the food we eat. Our digestion depends on living organisms that reside within our bodies. As a system, our bodies contain ten times as many "foreign" cells as those that contain our genetic material.[159] In fact, all of the material that comprises your current body was once part of something else. A Lawrence Krauss has pointed out, every atom in your body was once inside a star that exploded.[160] In that sense, we are all reincarnated, or perhaps more accurately, recycled. In terms of time, which is only a concept of course, in each moment we each are simply one point in the endless cycles of life and death.

But what about cells that were built on the basis of our genetic material but changed into something else? In other words, what about cancer? Normally, cells have a self-destruct mechanism called a *telomere*. This is like a tiny clock that ticks off every time the cell divides, and once the telomere clock runs down, the cell dies. But cancer disables the mechanism that makes the clock run down, so the cells continue to regenerate themselves indefinitely.[161] Does this make cancer immortal? What about the individual whose cancer cells are kept alive in the laboratory? After all, those cells contain all of the genetic material that is specific to that individual. If the seed of self is contained in that material, can it be said to actually be that self, or is it indistinguishable from the broader pattern of life? In any case, the living cells of cancer continue to change and grow in order to stay alive.

KARMA CHAMELIONS

The *Buddha Science* concept of not-self perfectly aligns with these scientific findings. Not-self does not deny that there is a consciousness contained within a body, but it says that an autonomous and unchanging self is an illusion. Although in each instant, we are the result of causes and conditions from events that have happened in the past, we are also a product of our intent in each moment and perhaps of future events as well. We are never bound by the past. As Steve Hagan puts it in his excellent book, *Buddhism Plain & Simple*, "We die in each moment and again, in each moment, we are reborn."[162] In other words, individuals are merely an ever-changing part of a larger process. The correspondence to modern science is incredible when you consider that the concept of not-self was embraced by the Buddha before atomic theory was developed, before the discoveries of evolution and genetics, and before cellular structure was understood.

What about life as a whole? Is there some constant, unchanging essence to this universal quality of things we call alive? Because life is dependent on the environment for resources, the very presence of life creates chemical disequilibrium in that environment.[163] In the chapter on chaos, we have seen that many of the processes associated with life are complex and chaotic, so they never reach a steady state, or even repeat themselves exactly. If we had a detailed enough map we might be able to define the state of life at one particular period, but by the time we completed our measurements, that state would have changed. *Buddha Science* views life as a combination of physical and mental energies that are collectively called the five aggregates.[164] No matter what level we look at it, life cannot be defined as something static and unchanging. As *Buddha Science* has observed, everything is impermanent, every moment is unique.

Yet, once life developed on the earth, it continued and grew. Although 99 percent of all life species are now extinct,[165] evolution continues, and genes persist. In fact, scientists believe that the process of natural selection favors the perpetuation of genes over individual organisms. As David Barash says in *Buddhist Biology*, evolution is "nothing but a highly specified process of transmitting a kind of gene-based karma."[166] According to *Buddha Science*, there is no unchanging essence of life. But could genes actually play that role? Here we invoke the relativity of time. Although genes generally outlast the lives of individual organisms by continuing in their offspring, even genes evolve, mutate, and die.

A particularly dramatic example of gene evolution involves the Antarctic ice fish. This fish was discovered during a 1927 Norwegian expedition. The blood of this fish literally runs as cold as ice. Not only does the blood contain a natural antifreeze, it is unique to the world of vertebrates in that it has no hemoglobin, so its blood is crystal clear.[167] The process that caused this unusual creature is a startling example of natural gene splicing, illustrating the dynamic and fluid nature of the genetic process, as well as the intimate interconnection between genes themselves.

So how did this all come about? How did life emerge and evolve? The Buddha reportedly declined to speculate on the origin of the universe or of living beings, dismissing such questions as unanswerable.[168] Science itself has struggled to answer the questions, invoking a number of possible mechanisms to explain the emergence of life on earth. These include such mechanisms as:

- Lightning striking the primordial soup (the Frankenstein effect)

- Deep sea vents that spew warm, nutrient-rich water into the oceans (ocean's womb)

- Ocean ice that protected the fragile components from ultraviolet rays, allowing the time necessary to develop life (you're cold as ice)

- Clay as a catalyst to organize molecules into life structures (an arranged marriage)

- Life arriving from other planets (hitchhikers from the galaxy)[169]

Each of these theories depends on random factors to one extent or another.

Jeremy England, a researcher at MIT, recently developed a mathematical model based on established physics. This model suggests the origin of life may be less surprising and mysterious than the other models might indicate. Basically, the model is based on the premise that a group of atoms will restructure itself when it is driven by an external source of energy.[170] In his view, certain types of inanimate material are similar to life in the way they "eat energy" from the environment.[171] This may provide one tool for understanding the process that leads to the emergence of life from inanimate matter. Rather than replacing Darwinian evolution, the mechanism Professor England describes could provide a look at one of the details in the process that drives evolution in general. In this view, life is an emergent process of a complex system that includes groups of atoms and an external source of energy. As with any new idea that challenges the conventional wisdom, this model is being met with mixed reactions in the scientific community. Only time will tell if this new map looks more like the terrain than those that came before.

Evolution itself is well documented through the fossil record. The geological record shows evidence that life as we define it has been around for 3.6 billion years, but it wasn't until almost a billion years later that life began to have a significant impact on the earth. This impact began when cyanobacteria emerged, a life form that creates oxygen through the process of photosynthesis. This began a slow build-up of oxygen in the atmosphere that would accelerate over the eons and pave the way for the emergence of more complex life forms. Animals emerged around 600 million years ago, early humans emerged at around four million years ago, and the modern human form has only existed for around 200,000 years, a slight fraction of the time since life first emerged.

Evolution itself is highly compatible with *Buddha Science*. As we have seen, in *Buddha Science* every being is based on the causes and conditions that brought it about. This can be considered an accurate general description of the evolutionary process. Parts of our anatomy can be traced directly through ancestral creatures that existed hundreds of millions of years ago.[172] The biological, chemical, and physical interconnection of life can be considered a form of reincarnation.[173] The *Buddha Science* observation of interconnection is clearly demonstrated, not only through our connection to other life, but also to the universe itself.[174] Buddhism embraces impermanence, also a basic element of evolution.[175] In fact, as we will see in the next chapter, even the Buddha's diagnosis of suffering in the world can be explained through the science of evolutionary psychology.

OUT, STANDING IN THE FIELD

Life does not exist in a vacuum, of course, but in relation to its environment. The frogs and fishes in my yard would not exist without the pond that surrounds them. The immediate connection we have to our environment is studied through the science of ecology, the examination of life and its environment as a system. Ecology is based on the scientific understanding that organisms and their environment are inextricably interconnected. Not only does the environment affect the propagation, life, and distribution of organisms, but also organisms affect each other and their environmental surroundings. We have already seen that the emergence of cyanobacteria changed the entire atmosphere of earth. Although this process took nearly a billion years, living communities have impacted their environments in ways both large and small since life began.

Communities are critical to the survival of species and therefore of genes, but as communities impact their surroundings, the surroundings also impact the communities in a never-ending dynamic. Living communities can be chaotic, never settling into a perfectly repeatable pattern. Forest succession is a good example of this dynamic interchange between communities and their environment. Moss and lichen may begin to take hold in a rocky field, breaking down the rocks, forming soils, and increasing the ability of the soil to hold moisture. This allows grasses, first annuals and then perennials, to grow. Shrubs begin to appear, and then hardy, fast-growing trees such as pines take hold. These may be followed by a succession of softer, faster-growing deciduous trees, and finally hardwoods that mature over time. Where large older trees eventually die, another succession will take place in the resulting open areas. Throughout this process, bacteria are breaking down organic materials, worms are working the soil, bears are doing what bears do in the woods, and various creatures are helping to pollinate, create burrows, and add to the complexity of the system. As each new event causes change to the system, the system dynamically adjusts, which causes new events in an ongoing process of birth, death, change, and renewal.

Plant communities seem to follow genetically predetermined patterns, but can they also exhibit the emergent behavior that is a characteristic of complex systems? Little *Shop of Horrors* aside, there are plants that challenge our ideas about emergent behavior in the plant world. One enigmatic example of emergent behavior is that of slime molds. Although they are one-celled organisms, when their food supply is threatened they are capable of banding together to form a structure that creates spores, allowing the mold to be transported through the air or on the bodies of insects and animals, to other locations.

Forests themselves share resources through the underground network of root systems. Professor Suzanne Simard of the University of British Columbia describes this as a neural network that includes trees and fungus acting in connection to create a communication network that makes the forest resilient to drought and fire.[176] Nutrients and resources, such as carbon dioxide, are shared between every tree in this network, no matter what species they are. This network helps the forest to survive through conditions that would kill individual trees, allowing it to continue manufacturing the oxygen that we, as humans, need to survive. This is an emergent property of forests that depends on a certain level of complexity, which can be destroyed by activities such as clear-cutting. As a Buddha Scientist might say, there is a direct karmic connection between the way we treat the forests and our own lives.

Plants can achieve movement, and therefore propagation, through the production of spores or seeds that have various clever mechanisms for transportation. However, these mechanisms are in large part passive—they rely on wind or water, insects or animals to help them insure survival of their genes. Animal and insect populations, on the other hand, have the advantage of intentional movement. This provides more control, but it also allows fierce competition for resources, as animals and insects have the ability to rapidly move into territory already occupied by others and aggressively commandeer the resources in that area. This can result in extremely rapid and dramatic change in the environment.

Often, this happens as a result of the species being relocated for one reason or another. In their native habitat, a given species may be perfectly innocuous, or even beneficial. The species is in balance with its ecosystem because it has natural competition from other species within that system. But when the species enters a new ecosystem, it may become invasive due to lack of competition. A prime example of such massive change through transport is the Emerald ash borer. These insects, native to Asia, were discovered in the United States in 2002, and have worked their way into at least 22 states, destroying tens of millions of ash trees. Generally, these insects do not have the ability to fly far from their home, but are spread to new areas by people unknowingly transporting them in firewood, wood chips, and nursery stock.[177]

Some animals have dramatic impacts on their native environment as well. Elephants, for instance, roam across vast territory, destroying trees in their quest for food. Yet when they have the amount of territory they need, the system has time to recover so it can achieve a dynamic balance. It is only when the territory is limited by competition due to human encroachment that the environment will be dramatically altered. Goats will overgraze and overpopulate in some areas if left unchecked. Some areas in Australia have been completely denuded where these animals were introduced to support human habitation.[178]

Though human beings, as a species, tend to judge these changes as good or bad for the environment, the environment itself doesn't "care." It simply adjusts to find a new balance, or changes according to whatever new dynamic presents itself. The environment acts without intention. This observation is elegantly expressed in the words of Buddhist monk and Vietnamese Dhyana Master Huong Hai:[179]

> *A silver bird*
> *flies over the autumn lake.*
> *When it has passed,*
> *the lake's surface does not try*
> *to hold on to the image of the bird.*

Humans inject intention into the equation. It is not possible, at least at this point, for humans to destroy or save the planet earth. It existed for billions of years before we existed, and will likely continue to exist for billions more. Our actions can change the current state of the earth, and the ability for any species to survive, including our own species, but we cannot know what effect our efforts have on the long term dynamic of the earth or the universe. The idea that we can "save the planet" is purely a human concept; which, ironically, was developed by the very creatures that have changed the planet so dramatically. So it's pretty arrogant to think we can save the planet, or destroy it for that matter. We might be able to render it unfit for human habitation, but if we do that, earth has billions of years to create new patterns of life. This is not to say that I believe it's ok to trash the place. In fact, just the opposite. This interconnected web of life that is the planet earth is so miraculous and complex that we should try to avoid any intentional or unnecessary action that could disrupt that miracle. But I also think that life has its own pattern, and that pattern will prevail over our limited efforts to affect it. The planet does just fine all by itself. Like my pond, the earth's dynamic processes would work better if the beings that think they can save it would just leave it alone.

Yet we can ask questions about our impact on the planet because we evolved a unique trait that allows us to compete with all other animals for earth's resources. We evolved a degree of intelligence that allows us to use found objects as tools, to invent new tools, and to radically alter our environment in order to raise our own food. As a species, our intelligence allowed us to build new types of community, as we will discuss in a few moments. But before that, we need to ask some questions.

What is intelligence? As we have seen before, the limitations of language fail us. Intelligence is a term that has been defined in many ways without general agreement. One useful definition based on composite information from many authors includes the following three abilities:[180]

1. The ability to learn through experience, education, and training

2. The ability to pose problems, by recognizing there is an issue and forming a problem concept

3. The ability to solve problems by fashioning a solution to the problem as conceived

Does intelligence require a large brain? We have already seen that slime molds can develop a community to build structures for passive transport of spores, but they also can exhibit behavior that looks very much like intelligence. Solving a maze, for instance, would seem to require a certain level of reasoning, but slime molds have shown the ability to do just that.[181] Slime molds are one-celled organisms that have no brain or nervous system, yet when their food supply runs low, they can combine to move toward food even through complex mazes in the laboratory. In fact, they are so resourceful at designing the most efficient routes between food morsels, that some researchers have proposed using them to help plan future road construction. [182]

Is this apparent intelligence an emergent behavior as described by complexity studies? Do these single-celled organisms communicate through entanglement as described by sub-atomic physicists, this time operating on a macro scale? Does this truly involve the three abilities described above, or is it simply some mechanical honing mechanism that drives them toward food? Slime molds are an enigmatic example of apparently intelligent behavior from an organism that doesn't appear to possess the equipment for such behavior. However, the confusion may not be in the behavior itself, but in our concept of what intelligent behavior looks like.

By now we shouldn't be surprised that our concept of intelligence is not Reality, yet it is our intelligence that allows us to form the concept of intelligence in the first place. Before we get hopelessly enmeshed a conundrum of self-referential logic, we need to leave the enigmatic example of slime molds behind, and revisit the question that evolutionary advantage provided through our human ability to learn, and to identify and solve problems. Although many animals evolved intelligence to varying degrees, the human animal has developed that ability to the extent that it includes complex language, religion, and elaborately built environments. Of course, the ability to articulate language is also a function of the particular vocal mechanisms that we have, and our opposable thumbs allow us to create and write symbols that express complex concepts. So it is the combination of various physical abilities combined with our intelligence that has allowed human beings to develop our current state of dominance, ecologically speaking.

From an evolutionary perspective, these abilities more than compensate for human beings' lack of speed, strength, and endurance, allowing us to out-compete other animals for earth's resources. We have the ability to modify our environment on a massive scale. It remains an open question as to whether we will use this ability to transform the earth so radically that we cannot survive as a species. Our survival is no longer a matter of being the fittest in terms of competition-based survival, but a matter of our willingness and ability to intentionally work together to find solutions to the problems that face us as members of the world we live in.

So the paradigm has changed. We have the ability to see and understand the interconnectivity of all living beings and the environment. We have the ability to see that there are far-reaching effects of the decisions we make and the actions we take. We have the ability to realize that we no longer need to compete with other animals for scarce resources, and that working in cooperation with other humans would be more conducive to perpetuating ourselves as a species.

Yet we persist in competing for resources through war between different factions within our own species. We divide ourselves along racial, ethnic, and national lines. *Buddha Science* says all of these dividing lines are illusions, and these illusions are the source of human suffering. Why do we persist in believing these illusions, and possibly destroying the very foundations of our life and our societies? Could it be that 2600 years ago, the Buddha saw that the paradigm has changed, that the evolutionary forces that drive us have caused a disconnect between the way we react to our environment and our ability to find happiness? Was the Buddha advocating a rebellion against our basic nature in order to helps us find a more peaceful existence in accordance with the new paradigm?

To explore these questions, we need to look more closely at the way our minds work, not only through the observations of *Buddha Science*, but also through the lens of psychology and neuroscience. That will lead us into asking about consciousness itself. We will tackle these topics in the next chapter, on the science of mind.

CHAPTER EIGHT
SCIENCE OF THE MIND

Let us understand, once for all, that the ethical progress of society depends, not on imitating the cosmic process, still less in running away from it—but in combating it.
—Thomas Henry Huxley

TO BE OR NOT TO BE AN ANIMAL?

Ever since Darwin began the evolution revolution, we have struggled to apply or reject "survival of the fittest" to the progress of society. Thomas Huxley, a proponent of the theory of evolution, wrote the words above to underscore his concern that the theory was being used to defend unethical behavior,[183] or "social Darwinism," as we would call it today. The study of life, how it emerged and what it is, inevitably leads to questions of nature versus nurture, of overcoming versus embracing our animal instincts. Personally, it's been quite a while since I slung my dinner over my shoulder and loped back to my man cave to eat it. I overcame that particular animal instinct when I got married. At any rate, the battle against our animal nature, and the moral dilemma it poses, is recognized by science and *Buddha Science* alike.

The middle way of Buddhism escapes this dilemma by taking the approach that if we can see and recognize our true (Buddha) nature, we'll understand that our animal nature does not give us a true picture of Reality. We'll see that our thoughts and ideas are merely concepts, a result of causes and conditions. And, as we have already discussed, we will see that our belief in a constant, unchanging self is a mistaken concept. If we do not cling to these concepts as if they are Reality, they will lose their power over us, and we will no longer be subject to the suffering that our clinging causes.

Science, on the other hand, combats this dilemma with the intellect. The scientific approach says that if we can study and understand how our minds work, we can combat our animal nature with reason and logic. In a sense, science is exactly the opposite of the *Buddha Science* approach, because it relies on concepts to gain this understanding, and this entails the belief that these concepts represent at least some aspect of Reality.

This approach follows from the scientific idea that we can separate ourselves from our objects of study, even if that object is ourselves. In this study of our own minds, the distinction between science and *Buddha Science* comes into sharp focus, because we are both the subject and object of study. Is it even possible to be objective about ourselves and the working of our own minds? Or do our minds function at such an unconscious and instinctual level that we can never remove the "log from our own eye" in order to see clearly? If the log is invisible to us, is it still possible to remove ourselves sufficiently to see it in the eyes of others?

Traditional psychology depends on the therapist maintaining distance from her patient in order to clearly see the demons he needs to exorcise. *Buddha Science* says the way to free another is to help him see the illusions that his basic human nature causes him to believe are Reality. For *Buddha Science* then, it's not about exorcising demons, it's about simply recognizing that they are demons we all think we have, but they don't actually exist. Is it possible that these two distinctly different approaches can find the same truth? The modern study of evolutionary psychology helps to explain how our unconscious thoughts and motivations may have come about, and helps to bridge the gap between science and *Buddha Science*. One approach to understanding the similarities comes through looking at the four noble truths that we mentioned in Chapter 3. As a reminder, these four noble truths can be stated as follows:

- First noble truth—Life is unsatisfactory, and causes suffering and anxiety

- Second noble truth—We suffer because all things are impermanent, yet we crave pleasant things and push away unpleasant things

- Third noble truth—We can escape the suffering

- Fourth noble truth—Outlines a path that leads to escape from suffering

Robert Wright, an American journalist, scholar, and author of many best-selling books about science, has looked in detail at the issue, developing specific ideas about how evolutionary psychology relates to the four noble truths espoused by the Buddha.[184] The next few paragraphs summarize some of the work that he presented in an online Coursera course entitled Buddhism and Modern Psychology during the spring of 2014.

Buddha Science says that the reason we cling and therefore suffer is that we fail to see the world clearly. We do not see the impermanence of pleasure. We do not see how quickly it fades and we do not see how much we will suffer when the pleasure does fade. Professor Wright uses an example of infatuation with another person. If you are infatuated, you may hold a besotted view of the person as perfect, and believe that you have finally found your one true love. You convince yourself that if you are just able to get into a relationship with that person, everything in life will be better, probably forever. In fact, you constantly convince yourself that the next pleasant thing will be much better—that next piece of cake, the new job, the next sexual encounter. But when you finish that first piece of cake, the satisfaction doesn't last and you are ready for the next one. If you get the job you have been trying to land, you soon find out that it's not as perfect as you expected it to be. This constant dissatisfaction with your lot in life is what the Buddha was describing in the second noble truth.

Dopamine is a chemical in our brains that correlates to pleasure. Primate's brains are structured in a similar way to ours, so measurement of dopamine levels can be used as a scientific way to determine the amount of pleasure a primate is experiencing. Referring to a study of dopamine levels in monkeys,[185] Professor Wright made the following comments:

> *They gave a little fruit juice to a monkey and what they saw was a dopamine spike that lasted about a third of a second. Assuming that spike is correlated with pleasure, that's pretty brief pleasure. If the monkey's condition is similar to that of humans, it is all the more reason to look at natural selection as a way that pleasure does evaporate. Why does natural selection "design" brains where pleasure is so fleeting? And why don't we get the picture in our everyday lives that the pleasure is going to evaporate so rapidly?*
>
> *Although natural selection is not a conscious designer, we can ask the following as a thought experiment. If we were designing organisms, how would we design their brains in order to get their genes into the next generation? Things like eating, sexual encounters, and elevating their social status will help them perpetuate their genes. There are three principles of design that would make sense if you want animals to reach these goals:*
>
> *1. Deliver some pleasure after the animal reaches important goals*

2. *Make the pleasure evaporate shortly thereafter so the animal will want it again*

3. *Make the animal focus more on the pleasure that is to come than the ensuing evaporation*

The Buddha said that pleasure tends to evaporate, it leaves us unsatisfied, and we do not get the picture—we focus on the pleasure, not on how fleeting it is. These all make sense in terms of natural selection.

Getting back to the monkey study, the experimenters trained the monkey to turn on a light that would release the fruit juice. They found that a dopamine spike happened when the monkey turned on the light, but not when it drank the juice, so the anticipation of the reward was what appeared to cause the greatest pleasure. When the experiment was altered to eliminate the fruit juice reward, the dopamine spike still occurred when the monkey turned on the light, but dopamine levels actually dropped at the moment the fruit juice would otherwise have appeared. This could represent the letdown of unfulfilled anticipation.

Professor Wright summarized by concluding that there is clear correspondence between the way you would expect natural selection to "design" a brain and the first two noble truths taught by the Buddha. He also observed that natural selection doesn't "care" if we see the world clearly, or if we are happy. As we have seen in the previous chapter, the primary "goal" of natural selection is to perpetuate genes into the future, rather than support the needs of any given individual. This is the basic problem of suffering that the Buddha sought to address, and in the view of Professor Wright, Buddhism is therefore a rebellion against the illusions that are built into us by natural selection.

The next two noble truths offer the potential to escape from this suffering, whether you think of this as rebellion or simply seeing things more clearly. As we have said, the third noble truth says there is a way out of our dilemma, and the fourth noble truth describes the way out, which the Buddha called the eightfold path. Although this path is potentially a subject of lifelong study, I have selected the following list of explanations from the *Religion for Dummies Cheat Sheet* [186] by Rabbi Marc Gellman and Monsignor Thomas Hartman, because it very effectively presents an explanation of the path in simple, modern language. It is presented here with minor modifications:

- Right understanding: Understanding that the four noble truths are noble and true.

- Right thought: Determining and resolving to practice Buddhist faith.

- Right speech: Avoiding slander, gossip, lying, and all forms of untrue and abusive speech.

- Right conduct: Adhering to the idea of nonviolence, as well as refraining from any form of stealing or sexual impropriety.

- Right means of making a living: Not slaughtering animals or working at jobs that force you to violate others.

- Right mental attitude or effort: Avoiding negative thoughts and emotions, such as anger and jealousy.

- Right mindfulness: Having a clear sense of one's mental state and bodily health and feelings.

- Right concentration: Using meditation to reach the highest level of enlightenment.

This path outlines a set of practices that leads to respect for all life, caring for others, and seeing the interconnectivity of all beings so as to avoid causing harm in the world. So, while the path may be a rebellion against the old paradigm of competition-based survival, it could also be considered a useful approach to supporting, as Huxley put it, the ethical progress of society. Although the Buddha lived many years before Darwin, could he have sensed that our basic dilemma was failure to move from the old paradigm of natural selection to the new paradigm of social order?

The Buddha did not present the eightfold path as a way to perpetuate society, but as a way to eliminate suffering in the world. It was not a formula to eliminate physical pain and illness, but to eliminate the sense of dissatisfaction we experience when life does not meet our mental image of what it should be. As the story goes, when a man came to the Buddha with a litany of problems, the Buddha said, "You will always have 83 problems, and I cannot help with those. But perhaps I can help you with the 84th problem." When the man asked what the 84th problem was, the Buddha replied, "You want to not have any problems."[187] So our view of a perfect life is an illusion, and *Buddha Science* seeks to address the suffering that comes from clinging to that illusion. In fact, we cling so hard that our clinging becomes craving for life to be a certain way, and if it isn't, we suffer.

According to teachings attributed to the Buddha in the original Pali texts, our craving takes three forms. One form is craving for sense pleasures. The above discussions regarding the work of Robert Wright is one way that modern psychology helps to explain the basis for this craving. The other two forms of craving described by the Buddha are craving for existence, and craving for non-existence.

What is craving for existence? In general, it is the need to cling to our belief in an unchanging and permanent self. As we have previously discussed, *Buddha Science* holds this belief to be an illusion. The need manifests itself through an intense craving to become something, to make a difference, to create something that lasts beyond us. This craving is similar to the Freudian concept of Eros. Eros is our basic instinct—a desire for preservation, which seeks success, growth, and connection.[188] We experience tremendous suffering because we are unable to reconcile this basic desire with the Buddhist and biological view of the individual as an illusion.

Why is it that we continue to believe in an unchanging self and in our separation from the world? These two beliefs, after all, are what drive the craving for existence. Although not everyone would agree, the modular model of the brain may suggest some answers. This model posits that we are controlled by a large number of specialized modules that evolved through natural selection for very specific purposes, and when the situation demands, the modules that will help us (our genes) survive are activated. These modules may be designed for such things as selecting a mate and morally condemning others.[189] Selecting a mate helps us perpetuate our "selves," while condemning others supports our belief in separation. The modular model also suggests that the self does not exist as a tiny CEO that controls us, and that even what we think, we believe may be controlled by unconscious modules.[190]

Some of these modules talk to each other, and some do not. So the model explains why we can hold conflicting views without seeing the conflict. In the introductory chapter of this book, I said that there were seven billion people on this planet and probably 20 billion views of Reality, and I promised to explain later. Well, now is later, and that is the explanation. It's likely that most of us have a number of views of Reality, based on what is most advantageous to believe at any given time. It's also reasonably likely that we're not even aware of it, because the different views we have don't talk to each other. Earlier on, I suggested that I could use the Planck wall as an excuse when my wife accuses me of not listening. Instead, maybe I should try saying, "I did listen, but it's not my fault. My currently active module wasn't paying attention to the one that heard you." I'll bet that will convince her.

The self that we consciously think of is not only constantly changing and inconsistent, but it represents a mere fraction of the complex processes that drive our actions, thoughts, and beliefs. The modular model suggests that, whether we are consciously aware of it or not, we are genetically programmed to perform actions and hold beliefs in concepts that will help our genes survive. This modern concept of the brain, then, supports the observation of *Buddha Science* that we crave existence.

One weak link in the modular model, from my point of view, is that some proponents of the model try to explain conscious awareness of our actions by postulating that individual modules "kick into gear" based on the situation that occurs, and once they are activated they take conscious control of our actions. This would mean that rather than one CEO self, we have a bunch of "little selves" that are each designed to do some specific task,[191] and that some of these modules must be involved in conscious decision-making.[192] But it seems to me that this just compounds the problem. If this concept is correct, we need to explain how dozens of modules evolved conscious awareness, rather than just one. A little later in this chapter, I will be providing an alternative explanation that I believe answers this problem.

It might seem as if craving for non-existence would be incompatible with this desire for preservation, but the two are closely related. This is the 84th problem described by the Buddha. While we may want to exist, *Buddha Science* says we want to do it only on our terms rather than experiencing the world as it is. We want to avoid physical discomfort and pain, the letdown we experience from unfulfilled expectations, and the isolation we feel when others criticize us. In other words, we want to escape the realities of life. This craving echoes the Freudian drive for Death, the part of us that seeks relief from disappointment and despair.[193] Some experts believe this drive is tied to the roots of depression.[194] The preponderance of drug production and use throughout the world is one result of a basic human drive to escape from the pain of existence. The United Nations Office on Drugs and Crime estimates that worldwide there are over 50 million users of opiates and amphetamine-type stimulants, and over 180 million cannabis users. In addition, the types of new psychoactive substances have increased by 50 percent from 2009 through 2012.[195]

Addiction can also take the form of gambling, theft, sex, or other behaviors. It doesn't include eating massive amounts of dark chocolate, of course. Although we typically draw a dividing line between addiction and things like a single-minded focus on work or on the creative fugues of an artist, if such things interfere with life and relationships, they can have an addiction-like effect on our lives. Even an overdependence on relationships can be considered an addiction. Although experts disagree on the reasons for addiction, even with physical addictions there is typically a strong psychological component, so people who are prone to addiction frequently switch from one type of addictive behavior to another. [196]

In fact, addiction is a name that can apply to a whole range of behaviors that are related to all three types of craving discussed by the Buddha. As an example, while an artist may start out searching for the sensual pleasure of creation, he may become obsessed with building immortality through his creative works. When he realizes the folly of this obsession, he may seek to eliminate existence through drinking, or even suicide. The story of Vincent Van Gogh is a well-known example of an artist who became so obsessed with creation that he often couldn't see how much his brother Theo loved and supported him. There are many others we know of in contemporary society who seem to have had everything they could ever want, but could not stand the pain of existence and so took their own lives. Although there are physical and genetic causes for mental illness, society tends to exacerbate the pain involved by shunning and further isolating those who suffer with these problems. We want to not have any problems, so we avoid those who do as if their struggle could somehow infect us. A Buddha Scientist would observe that each craving, and the pain that goes along with it, is rooted in the illusion that the person who experiences the craving is a separate self, removed from the rest of the world and dependent only on oneself. Unfortunately, since most of us carry that illusion, we can be unable to help those who experience pain more acutely due to mental illness. We can unwittingly contribute to making the environment unpleasant for them.

In a study conducted in 1978, researchers concluded that unpleasant environmental conditions appeared to drive rats to become addicted to morphine.[197] One cause of addiction, then, may be our built-in craving for non-existence coupled with an unpleasant environment that activates this craving. Our environment includes our family, our training, our physical environment, our economic situation, our society, and the mental environment that relates to all of these factors. Any or all of these factors could cause us to experience an unpleasant environment. Although we may not be able to avoid the chemical and genetic reasons for mental illness, we all contribute to the environment that is experienced by those who suffer from it.

Modern psychology utilizes various symptoms to make a determination of psychological addiction,[198] and this supports the observation that such behavior is part of a continuum—there are no bright lines between addiction and "normal" behavior. Clearly, if a person gambles away their entire life savings after 30 years of work, this is extremely unusual behavior. But is it addictive? If we discover that they had just been diagnosed with a rare, untreatable disease and given 24 hours to live, although we might think there were better ways to spend their last moments on earth, this could be considered a normal reaction to the circumstances. A person who has to live with intense chronic physical pain may develop a physical and psychological addiction to painkillers, but is that necessarily worse than the alternative of debilitating pain? *Buddha Science* and modern psychology would both agree that craving is basic to all humans, and the resulting behavior is a continuum rather than an on/off switch.

Yet in our modern world, we treat addiction as a discrete condition. Once some magical threshold is reached, we label the person as an addictive personality and set up treatment programs that include counseling, medication, self-help, support groups, and so forth. Some Buddhist facilities offer successful drug rehab programs that seek to change the mental formations that lead to addiction, similar to the twelve-step programs based on the one for alcoholics founded by Bill Watson and Dr. Robert Smith in 1935.

In fact, it is useful to look closely at the similarities between twelve-step programs that have become the treatment of choice for addiction in the US and the eightfold path outlined by the Buddha. This may seem like an odd comparison, because twelve-step programs are designed to address a specific behavioral problem, while the eightfold path addresses the existential suffering that we all experience as conscious beings. But after all, addiction is simply an extreme form of the basic human problem as described in the second noble truth, and the eightfold path is a program to free us from the suffering associated with that problem.

Before looking at the similarities, though, we need to make one thing completely clear—*Buddha Science* flatly rejects one basic assumption behind typical twelve-step programs. Twelve-step programs start from the assumption that there is a self that needs to be healed, and that healing can only take place through surrender to a Higher Power that is outside that self. As we have discussed, *Buddha Science* observes that there is no separate, unchangeable self, so the idea of invoking anything "outside" of self is meaningless. In the Buddhist view, we simply have to drop the illusion of separation, and we will see that there is nothing to heal. A second distinction is that although the Buddhist approach is called a "path," there is no sequential order to the eight principles, whereas the twelve steps are more of a sequential unfolding toward recovery. So with this understanding of some key differences, we can observe that many of actions prescribed by these approaches to alleviate existential suffering are similar.

One striking similarity in these programs is, in fact, related to the difference mentioned in the last paragraph. That similarity is the general idea of surrender. While twelve-step programs advocate surrender to a higher power, *Buddha Science* advocates surrendering the idea of a separate, independent self.

In both cases, following the prescribed "path" is made easier through the support of a group of like-minded individuals. With Buddhism, the sangha is a key component of the practice, and with twelve-step programs, regular attendance at meetings is critical to success. And of course, one of the basic observations we have been discussing throughout this book is interconnection, so in fact, we never have to suffer alone if we can stay mindful of this truth.

The first teaching on the eightfold path is right understanding, which includes the understanding that there is no self in charge. Reality unfolds on its own based on causes and conditions that we have no control over. This is similar to the first of the twelve steps—the admission that the self is powerless to control behavior through will power alone. The eightfold path advocates acting from love and compassion, while the twelve-step program advocates making amends to all those we have harmed and admitting our faults to others. The eightfold path advocates using full and diligent effort in following the path, while the twelve-step program tells us to continue taking personal inventory of our behavior and promptly working to correct it. Finally, twelve-step programs promote continued seeking through prayer and meditation, and to seek spiritual awakening. The eightfold path is realized through mindfulness, which involves focus and concentration, aided through the process of meditation. And, of course, the release that comes from following the eightfold path is awakening in the Buddhist sense.

But once again, our terminology gets in the way. What, exactly, do we mean by awakening, or enlightenment? Robert Wright asks the question: "Is enlightenment something like what a psychologist might say your consciousness would be like if we stripped it of all the misperceptions and delusions that were built into us by natural selection?"[199] We want to take a closer look at what we have discovered about the working of the brain, mind, and consciousness, to see if these studies might be pointing to the same moon as *Buddha Science*. But first, we have to revisit the question of subjectivity versus objectivity. As we have seen, evolutionary psychology shows us that we may have built-in bias when it comes to our conscious pursuits, and when it comes to science of the mind, we are both subject and object. So we ask once again, is it possible to actually be objective when turning the microscope on ourselves?

In previous chapters, we began from the scientific perspective of objective measure and looked at how the observations of *Buddha Science* related to those findings. You may have noticed that so far in this chapter, we have taken the opposite approach. We have started from the Four Noble Truths of the Buddha and asked if they were consistent with psychological findings. We have taken this approach for three reasons.

- First, psychology and the related sciences study the mind, which itself is a concept. Although the mind seems to have some relationship to the brain, we do not fully understand that relationship, so we cannot specifically tie the object of study to measurable physical phenomena

- Second, the system we are discussing is self-referential. In other words, we are using our minds to study our minds. This leads to the concern that objective measure may not be possible.

- Third, psychology deals with nuances of feeling and behavior that often cannot be quantified, modeled, or tested in a laboratory setting. Psychological research methods include such things as our moral judgments of others, as opposed to the more definitive measures possible with the physical or life sciences, such as relative velocities or population counts.

Is there a way to get at this more directly, a way to look at some physical system to understand the nature of the brain, the mind, and perhaps of consciousness itself? Such studies are the focus of neuroscience, which looks at the physical properties of the brain. This provides a starting point to understanding how the physical system of the brain may be related to the mind and to consciousness itself.

BRAINS AND MINDS AND SELF, OH MY!

So first of all, what do we mean by brain, mind, and consciousness? Brain is the easiest of the three. It can be defined as the organ of soft nervous tissue contained in the skull of vertebrates that functions as the coordinating center of sensation and intellectual and nervous activity.[200]

Definitions of the mind are more variable, so we'll use one that helps us distinguish it clearly from the other two terms we will be working with. For the term *mind*, we'll start from definition number two listed on dictionary. com: the totality of conscious and unconscious mental processes and activity.[201] This definition is somewhat problematic in that it is self-referential. The mind is mental activity, but mental activity can only occur in the mind. So we will take the reasonable leap to connect mind with brain and change one word for our working definition of mind: the totality of conscious and unconscious brain processes and activity. Although strictly speaking, this is inconsistent with Buddhist terminology, it would be easy to venture into the metaphysical here, so we'll attempt to limit our discussion to concepts that are potentially testable through scientific methods.

Consciousness is the most problematic of the three, because through the ages, mystics, philosophers, and scientists alike have used the term to describe very different concepts. As with the concept of the mind, we'll be using the term in a more limited, scientific sense to mean the property that allows us to be aware of the physical and mental activities that we are performing. Of course, you could argue that by this definition, most politicians wouldn't qualify as conscious, but we'll set personal opinions aside for the purpose of scientific integrity.

As to how consciousness arises, this is a much broader topic than we can tackle. Actually, I have no idea how to even begin to understand this question or articulate an answer, so we'll have to rely on those who are smarter than me. It was easy to find some. The most satisfying description, in my opinion, was articulated by the late Roger Wolcott Sperry in 1969. Sperry was involved in many studies that involved consciousness, including neurological surgery, psychology, computation, and neuroscience. He understood consciousness to be a special kind of property that is itself non-physical, but which emerges from physical systems, such as the brain, once it attains a certain level of activity.[202] So in this view, consciousness is an emergent property of the brain.

As a quick reminder of our discussion on complexity studies, the property of emergence can arise from certain complex systems. The brain clearly meets all of the criteria of such a system and is consistent with the modular model of the brain, in that the brain: 1) is clearly complex; 2) is not self-similar to the modules; and 3) Includes a self-referential feedback system.

B. Allen Wallace has said that the mind itself is an emergent process, though its source may not be purely physical.[203] Whether the mind purely arises from the brain or involves some non-physical source as well, this idea of an emergent property or process supports the modular model by addressing the problem of conscious modules, as we discussed earlier in this chapter. No single module needs to be conscious for the property of consciousness to emerge from the entire system. It is also consistent with *Buddha Science*. If our conscious awareness (self) is a dynamic, emergent process, it is in constant flux and cannot be found anywhere in particular. It is even possible, as some contemplatives describe, that consciousness exists everywhere and the mind is simply like a radio, a receiver that translates the signals of universal consciousness into the thing that we mistakenly think of as self.

I warned you that the question of consciousness is tricky, didn't I? That's because the term itself means so many things to different people. Mystics, philosophers, and metaphysicians have argued the question for eons. Much as I'd like to wave my magic wand here and put the issue to rest so we can get on to critical questions like, "What's for lunch?" or "What is the airspeed of an unladen swallow?"[204] it's just not going to happen in the next few pages. Probably the biggest problem with trying to understand consciousness is that the very idea is self-referential. Consciousness is both the subject and the object—we are looking for an objective measure of subjective phenomena[205] such as colors, imagery, word meanings, opinions, and conceptual models. If consciousness is defined as "the self in the act of knowing" as neurologist Antonio Damasio suggests,[206] and if "self" is a dynamic, emergent process rather than a static, permanent thing, then by the time we know what a specific moment of consciousness was, it has already changed.

If your head hasn't exploded by now, you're not paying attention. So to save both your head and mine, we'll contain our discussion by asking a limited question. How do we get from the purely physical processes in the brain to the emergent properties of mind and consciousness?

You'll note that by asking the question in this way, we're making the assumption that consciousness is somehow related to physical phenomena, rather than as some sort of universal property that exists independent of the physical universe. You might also notice that I'm exercising uncharacteristic caution and restraint here. That's because we are right against the edge of walking into the metaphysical realm, so we need to be very careful about limiting the current discussion to the empirical aspects of Buddhism as they relate to science. Toward the end of this chapter, we'll look at an alternative view of consciousness and its relationship to physical reality, but for now we'll maintain our orientation on the physical world.

We cannot deny the fact that consciousness exists. If it didn't, I wouldn't have been able to write this book and you wouldn't be able to read it. To invoke some non-physical, universal process for it is to assume that we can step outside our world and view non-dual Reality as a whole. But we are still blind to non-dual Reality, so we have no choice but to view the elephant from our vantage point within the physical world. In fact, this is consistent with our focus on *Buddha Science*, because scholars tell us the Buddha himself said very clearly that consciousness does not exist independently of matter, sensation, perception, and mental formations.[207] However, because mental formations are involved, consciousness does not arise purely from the brain, either. In Buddhism, the mental is never reduced to the physical.[208] This is an important point of departure, because if it cannot be tested in a physical sense, consciousness falls outside the purview of scientific study. But we do know three things: there are physical processes in the brain, consciousness exists, and these two things appear to be related to each other. What we don't know is whether one gives rise to the other, or if they are related in some other way.

One way to study consciousness is to try and create emergent behaviors within computers. A number of applications have been developed by using computer programs known as genetic algorithms, which are programs designed to evolve by natural selection. These programs set some rules for what constitutes successful adaptation, then they pass traits through generations of simulated critters, allowing for some variation in inherited traits, and see how the system evolves. Sankar K. Pal and Paul P. Wang have written a book that discusses application of this technique to pattern recognition,[209] which is one of the problems associates with interpreting satellite data. You might recall from Chapter 1 that this was one of the challenges that led me, years ago, to start thinking about ideas that have ended up in this book. For me, this is a clear example of interconnection. Perhaps some of the satellite processing brain cells I didn't use made their way into brains of people who were likely to employ them more productively.

Currently, a team of researchers at Michigan State University, led by Dr. Chris Adami, is using this type of computer simulation to try and create artificial consciousness. The team creates computer algorithms and robots that are given a few basic rules and a feedback mechanism. Then they turn their creations loose, see what type of behavior emerges, and how this behavior evolves.[210] If *Buddha Science* is correct, this approach may prove fruitful because it contains elements of the physical (the robots themselves), and mental formations (the programmed rules). On the other hand, we can ask if the appearance of conscious behavior in such a system is actual consciousness or simply sophisticated mimicry. For that matter, if you want to get science-fictiony about it, the only thing I know for sure is what consciousness feels like for me. Spock mind-melds aside, we can never directly experience the consciousness of others.

Let's look at this from the perspective of neuroscience, which is the study of the nervous system, including the brain. This system processes information, so it is often compared to a very fast computer. But there is one very significant difference. Conventional computers process data sequentially, one bit at a time. And software is typically written for this type of linear processing. Computation can speed up dramatically through parallel processing, in which a problem is broken down into independent parts and these parts are processed simultaneously. This parallel processing can be accomplished by combinations of single computers with more than one processor, networking a number of computers together, and with specialized hardware. In Chapter 5, we discussed quantum computers, which may have the potential to significantly accelerate processing speeds. But even with the fastest computers, each basic unit processes information in a linear sequence.

In contrast, the human brain works at a cellular level and is massively interconnected. It contains around 200 billion nerve cells (neurons) that are linked together by trillions of connections called synapses. The result is that the human brain could contain hundreds of trillions of connections that could operate simultaneously.[211] Recently, researchers used one of the fastest computers in the world (the Fujitsu-built K, ranked number one for speed in 2011) to simulate one second of the brain's neuronal network activity. It took 40 minutes. They concluded that the computer was capable of representing one percent of a human brain's neural activity.[212] So if the brain is a computer, it is a massively fast and powerful one.

Another factor that makes the brain different from computers stems from the basic structure of the brain itself. The brain is not a static piece of hardware, but an organic and interconnected cellular structure. Cells die and are replaced. As we have pointed out, virtually all of the protein in the brain is recycled every month. The synapses themselves, the connections that link everything together, grow, shrink, die off, morph, and are newly born in an ongoing dynamic process.[213] This dynamism may be part of what allows the brain itself to change in response to new conditions, a process known as neuroplasticity. Basically, neuroplasticity means that the brain can rewire itself based on experience, but there is no universally accepted definition beyond that. Researchers have found that this rewiring can occur at any age. The brain continually reorganizes and adjusts itself throughout life. Some connections are lost; others are newly formed, while new pathways can develop through existing tissue. A brain is not a thing, but a dynamic process.

On the other hand, the human brain does develop a broad organizational pattern that is generally stable throughout the life of its host body. In general, the higher mental functions such as planning, judgment, and creativity are associated with the front region in the brain, known as the frontal lobe. The region just behind that, which includes the cerebral cortex and a couple of other areas, controls motor functions, including eye movement, speech muscles, and voluntary muscle action. Behind that are the sensory areas, including the parietal lobe. These areas are associated with touch sensations, hearing, and language comprehension. In the very back lies the occipital lobe that involves sight and perception. The cerebellum is located at the base of the back, and is associated with coordination, balance, and posture. Moving forward from there, on the lower sides, are the temporal lobes that are associated with short-term memory and equilibrium. The main area for processing emotions, pain, and hunger is hidden away at the center of the brain. Interestingly enough, the region of the brain that is associated with olfactory sensation (smelling) is located at the front edge of this emotion processing area.

Although even this is a simplification of the actual structure, there won't be a quiz later. The point I'm trying to make is that science is beginning to develop an extremely detailed map of the brain, so through enhanced imaging techniques scientists are able to conduct research into topics that was never possible before. In fact, a recent study involving implanted electrodes was able to trace specific imagery and words to individual cells within the brain.[214] So what is this ability for detailed research revealing about the brain, mind, and consciousness itself?

First of all, we know that there is definitely a link between the brain and consciousness. Changes in the mental function, such as the use of mind-altering drugs, brain damage, and artificial stimulation of specific areas of the brain by researchers can have profound effects on a person's state of consciousness.[215] Contemporary scientists tend to distinguish three different states of consciousness: vigilance (being awake versus sleeping); attention (focus on something in particular); and conscious access (enough awareness of our attention that we can report it to others).[216] From imaging studies, scientists have discovered that a lot of preprocessing occurs in the brain before any of these three states occurs. From a purely subjective perspective, we never see how much preprocessing occurs before we become aware of something, since we can't be aware of what we're not aware of.

How do we know this preprocessing occurs? It is instructive to look at experiments conducted by Benjamin Libet, a researcher at the University of California. In one experiment, subjects were asked to simply flex their wrists whenever they felt like it. Using electrodes, Libet measured three things: "readiness potential," which we will roughly define as preparation for action, the start of movement, and a signal that Libet correlated with the conscious decision to act. What he found was that the "readiness potential" occurs almost one-third of a second before the decision to act.[217] In other words, even though we might think that simple wrist movement is a fully conscious act, the brain began processing the information a full one-third of a second before the subjects were consciously aware of making the decision. In fact, it appears that our brains can only consciously process one thing at a time, so there can be a further delay while multiple items sit in an unconscious mental buffer waiting to be processed.[218]

As an umpire for vintage baseball, I have experienced multiple examples of this buffer effect. One of my jobs as umpire is to call foul balls as quickly as possible so that the players who are already on base can make decisions about whether to run or stay put. On one particular occasion, the batter hit a ball that flew just along the third base line and curved to the left, so by the time it hit the ground it was out of bounds. I actually called it foul well before it hit the ground because I saw it curving. Although he accepted my call, I overheard the player telling his teammate that he thought it was a fair ball. He was correct that it started out heading toward fair, and of course he had first base to get to, so he caught the first glimpse of the ball flight, processed it, and moved on to more important business. My total focus was on the flight of the ball, so I had the time to process the curving trajectory more fully.

What kind of unconscious process is occurring that takes so long? After all, for a massively parallel processor like the brain, one-third of a second is about halfway to forever. Let's start by looking at what our sense organs actually record. Do we experience the world as it is? Our ears can hear only one-fifth of the range of frequencies that a mouse can hear. We can only discriminate between 1/25 of the different odors that can be detected by dogs.[219] We can only see a small portion of the total electromagnetic spectrum, and many animals can see in low light conditions that render us effectively blind. We cannot feel a mosquito land on our arm, detecting her only when she injects the saliva that causes us to itch. Humans have around 3,000 to 10,000 taste buds. Herbivores, such as cows, have around 25,000, and catfish have 175,000. Each taste bud is designed to detect one of five specific tastes: salty, sweet, bitter, sour, and savory (umami). Are other tastes possible if we had the right sensors? Who knows?

The point is: we have no idea what the physical world actually is, we only know how our bodies can sense it through the equipment we have available to us. Sure, we can design machines to detect phenomena beyond the range of our physical senses, but we design them on the basis of the things we can actually detect. Science suggests that dark matter and dark energy, combined, constitute 96% of the universe. But science has no idea how to directly detect these phenomena. Could it be that if we had the right physical senses we could either see dark matter and dark energy directly, or figure out how to design equipment to measure it?

Try this. Sit in a place with a lot of interesting visual items. Put your finger out at arm's length to the left, and focus on it as your sweep it around the scene. The unfocused background scene will seem to move smoothly behind the finger. Now hold your head perfectly still and try to scan the view with your eyes only, while focusing on what you are seeing. Do you see a smoothly moving picture, or a series of jumps? Without a moving target to track, it is extremely difficult to avoid seeing the scene with a series of jumping eye movements, called saccades. Yet, when you glance around without thinking about it, you don't notice this herky-jerky view of the world. This is because your brain has acted as video editor, filling in the gaps between the saccades.[220] This visual editing isn't limited to just smoothing out the effect of saccades, but can actually reinterpret images. In a number of experiments using goggles that invert or rotate vision, researchers have found that the brain quickly reprocesses the visual field to allow subjects to function normally, or even see a normal image even though the image is anything but normal. In fact, the eye is designed in such a way that the image delivered from the eye to the brain shows the world upside-down, but our brain automatically fixes that little design issue shortly after we are born.

So the conclusions we draw from the scientific findings about our senses is that, at a minimum, the various senses deliver merely a sample of the full richness of the world around us, pre-processed to make it coherent and comprehensible. A more sweeping view would be that what we see and experience is an illusion, a badly drawn map that we take to be Reality, but is merely a conceptualized version of the actual richness that surrounds us. By artificially stimulating specific regions of the brain, scientists have actually been able to artificially induce sensations that would normally be viewed as coming from our surroundings, such as burning, tingling, warmth, vertigo, and even levitation.[221] These feelings are concepts that exist in the brain, and are overlain onto our experience of the world.

As you have probably already figured out, this is highly consistent with the observations of *Buddha Science*, which describes the conceptual and nonconceptual aspects of mind. When we see a form, our mental concepts give meaning to the form, and we attribute that meaning to the object itself, rather than seeing it as a mental construct.[222] We confuse the map for the terrain. According to *Buddha Science*, this day-to-day conceptual view is the way we typically experience the world, but by realizing our Buddha nature, we can see the fundamental nature of the mind, which allows us to see the emptiness inherent in our observations. In other words, this realization of emptiness lets us "see" that much of what we experience is the concepts that we overlay onto our experience of the world, rather than inherent properties of the world itself.

So how do we become conscious of this experience? Can science help us determine when this unconscious processing results in conscious thought and action? Through a series of experiments, scientists like Stanislas Dehaene have been able to discover a number of brain activation patterns that occur when subjects become consciously aware of certain stimuli. For instance, when subjects became aware of words presented to them, the researchers saw an enormous increase in activity in the parietal and prefrontal lobes of the brain.[223] By measuring the strength of brain activity, the researchers were able to discover that as signals move through the brain from the visual processing centers toward the areas that do higher level processing; these signals can either lose or grow in strength.[224] Those that grow to a certain level seem to be associated with conscious perception. The researchers concluded that information from the sense organs may cause brain activity that sweeps forward and back across the brain like an avalanche, growing in strength until it reaches a threshold strength that triggers conscious perception.[225] If the activity never reaches that threshold level, we do not become consciously aware of it.

But consciousness is not what we think it is. Even when we become consciously aware of something, we may not fully understand what it is we are seeing or experiencing. Philosopher Dan Dennet presented a TED talk in 2003 in which he shows the audience various optical illusions and visual puzzles, demonstrating that what we think we see, we don't always see.[226] The same is true with thoughts, and association of feelings with objects. By the time we become consciously aware of phenomena, they have been filtered by our senses, sifted and preprocessed by our brain, so that even our motivations for making decisions may not be what we think they are. In other words, we don't actually understand how or why we think what we think. Our concept of our own consciousness is an illusion. We tend to believe that we are making conscious decisions, but what may seem to be conscious decision-making on our part is more like the justification we come up with to make sense of those decisions that have already been made subconsciously. Who or what is making those decisions? We really don't have any idea. There is no self in charge of the process. Once again, self is merely a concept that we develop to make sense of experience.

All of this points to the conclusion that there are two states of brain activity. The first, unconscious state, involves a collection of impressions, sensations, and images with unrealized potential that appear and fade constantly. Some of these coalesce and fade, and some trigger enough activity to enter into conscious awareness, which is the second state of brain activity. You could think of it as similar to feedback in a public address system. Either we have adjusted everything correctly and there is no feedback, or it builds to a horrible squeal, which definitely enters our consciousness. But when we become aware of any phenomenon, we do not become aware of all of the brain activity that led to that state of conscious awareness. The unrealized potential of our unconscious state has collapsed into the thing that now commands our conscious attention.[227]

If this process begins to sound familiar, it is because it is analogous to the quantum states that we discussed in Chapter 5. As we have discovered at the subatomic level, matter seems to exist in the state of probabilistic potential until an observation occurs, at which point the probabilities collapse into a particular state, either wave or particle. In the case of mental phenomena, the probabilistic state of unconsciousness collapses into a particular conscious concept.

Before concluding this chapter, I need to reiterate that the above discussions take a limited view of consciousness as something that emerges from physical phenomenon. This is a highly reductionist approach, based on what might be called material realism. I have taken this approach based on a limited view of consciousness as our awareness of actions and surroundings, because this is the aspect of consciousness that can be correlated with physical activity in the brain.

As we have pointed out, Buddhism does not reduce the mental to the physical. Correlation does not necessarily mean causation. For example, you may notice that pond ice thaws around the same time each year that the grass starts growing. This doesn't necessarily mean that thawing pond ice causes the grass to grow. As we suggested before, it is certainly possible that there is something out there called consciousness that exists independent of our brains, and that our brains are merely receptors of that consciousness, like a radio picking up electromagnetic waves from the air. In this view, when we see brain activity, it is our brain tuning to that particular radio station rather than actually generating the consciousness.

In his book *From Science to God,* [228] Peter Russell says consciousness can be likened to the light in a movie projector, which shines through the film of the mind to form images. When we see the images on the screen of a movie theatre, we don't think of them as light. In the same way, the people, places, and objects we perceive are not seen as projections of our consciousness, although that's exactly what they are. In this view, consciousness could be described as the pure light of being and the "ten thousand things" that we perceive are merely projections that our minds create to try and make sense of what we are actually experiencing.

But if the voices in our head come from some dark matter disc jockey or movie producer, we're a long way from being able to prove it. In fact, some scientists are coming out of the consciousness closet and declaring that we cannot fully understand our science without addressing the issue of consciousness in the larger sense—consciousness as a form of universal energy.

In his book, *The Self-Aware Universe,* physicist Amit Goswami holds the view that consciousness creates the physical world, and that the universe itself has conscious awareness. In other words awareness, not matter, is the ground of being. He proposes that the universe exists only in a state of formless potential until it is observed by conscious beings.[229] The state he describes can be considered equivalent to the pre-observation quantum state of subatomic phenomena, which collapses into a particular state once an observer is introduced. I'll come back to some of the ideas from Peter Russell and Amit Goswami in the final chapter.

Once again, we see that the two worlds of *Buddha Science* are consistent with the most current scientific findings and ideas. Ultimate non-dual Reality can only be vaguely seen and understood. Conventional reality includes the descriptive system we use to try and understand it, so it is meaningless to conceive of reality independent of the words we use to describe it.[230] That descriptive system is based on a language that requires us to collapse our concepts into subject/object duality. So at a minimum, reality as we experience it in a conventional sense is indeed created by our concept of it. As such, it limits our ability to see that consciousness is an integral part of existence.

We have developed various tools to help us come closer to understanding, but we will always be limited by the fact that we exist within Reality, subject to the limitations of our perspective. Over the millennia, we have developed many tools to help us understand bits and pieces of Reality, tools such as *Buddha Science*, formal scientific pursuits, philosophy, metaphysics, and religion. The best we can do is try to piece together the understanding gained from these various pursuits, and try to develop a better map, a map that may look a little more like the terrain of Reality than the one we have today.

In the concluding chapter, we will try to pull together what we have discussed here, and provide some thoughts and speculation about where we may be headed as we knit together these bits and pieces to try and answer the question we posed at the beginning of this book—What is Reality?

CHAPTER NINE
CONCLUSIONS

The problem is not the problem. The problem is your
attitude about the problem.
—Captain Jack Sparrow

SAY AGAIN?

The term "non-fiction book" is an oxymoron; there is no such thing. A book is only a map, and as we have seen, neither book nor map can ever be Reality. As a cartographer needs to determine who may be using her map and selects the things to include accordingly, this decision-making process is the same for an author. Science and Buddhism are both part of a dynamic, ever-changing terrain, and so even at the time this book was written, there was much more ongoing thought and research than it would ever be possible to capture in mere words. And much more has been done during the publication process.

In fact, any concept merely maps a portion of Reality. Borrowing on the words of Neil Gaiman, Reality is the only perfect map of Reality, and that map is so incomprehensible and indescribable that it is nearly useless. As Marcelo Gleiser says in his excellent book, *The Island of Knowledge*, "The map of what we call reality is an ever-changing mosaic of ideas."[231] The best we have is concepts. But concepts are not the problem; the problem is our attitude about concepts. The "real" problem is that we get so enamored of our concepts that we begin to believe they are Reality. Once we get a concept in our head about how things must be, we tend to order our lives, interpret what we see, and if we are a scientist, design our experiments around this concept. Every time we do something else to confirm this concept, it begins to harden into a given, and before we know it we can't view Reality in any other way. We confuse our concepts of what must be with our actual perceptions of what is.

In this chapter, with the understanding that we aren't talking about Reality but a way to understand some small portion of Reality, we get to have some fun with concepts. We have laid the foundation by discussing various scientific pursuits and how these pursuits relate to each other and to *Buddha Science*. We now will try to pull these ideas together to see if we can get a better glimpse of the moon of Reality. We'll do this by revisiting the broad observations introduced in Chapter 3 and pulling together the bits and pieces we have discussed throughout the book. To refresh your memory, these *Buddha Science* observations are interconnection, karma, impermanence, and illusion.

As we have seen, the idea of interconnection is simply that everything is connected to everything else. Another way to say it is that nothing exists without everything else. A tree does not exist without the seed, the sun, the soil, the process of photosynthesis, or the water it needs to survive and grow. The sun does not exist without fire, gravity, and the raw materials that formed eons ago. Each thing cannot exist without all of the other things, and that includes each and every one of us. We are connected to all of mankind, and our actions today affect not only our future experience in the world, but the future of us all.

Science confirms this *Buddha Science* observation of connectivity in many ways. The study of chaos introduced the butterfly effect, which postulates that very small changes in air movement caused by something as simple as the flap of a butterfly wing has the potential to cause major atmospheric disruptions on the other side of the world. Complexity studies involve the idea of emergence, whereby new properties emerge from the complex interaction of many simple processes. These properties cannot be attributed to any of the individual processes, but only to the whole system, which is dependent on the interconnectivity of all of the individual processes. Biology tells us that we are a collection of parts from all over the universe, and that our bodies are the result of eons of evolutionary development. Consciousness itself may be an emergent property resulting from the interconnection of the human brain, experience, and mental formations.

A basic premise of relativity theory is that the properties of objects, motion, and even time itself can only be defined relative to all other objects in the universe. And Einstein's famous equation, $E=mC^2$, says that energy and matter are interrelated. In the quantum world of physical science, there are indications that the interaction of fundamental particles may be closely related to consciousness itself. Through the phenomenon of quantum entanglement, physics has shown that subatomic particles can be instantaneously interconnected with other particles far removed from them in space.

Ecology shows us the complex interconnection of organisms with their environment. Not only is the organism dependent on the conditions of the environment, but the condition of the environment itself is inextricably connected to and dependent upon the organisms that exist within it. And these systems are also dependent on their past, including all of the evolutionary changes and interconnections that have occurred to give them their current configuration of genetics and functions. The human brain was developed through the same processes, and the modern concepts of evolutionary psychology demonstrate not only how strongly our past influences our thoughts and behavior, but also how closely these patterns correlate with the observations of *Buddha Science*.

Karma, the second principle of *Buddha Science*, has been described as moral interconnection, or alternatively as the principle of moral cause and effect. We know from the butterfly effect that minor actions can have dramatic consequences, and that is true in all complex feedback systems, including human interaction. We are, today, the product of processes, both seen and unseen, which have been going on for many years. The world we end up with tomorrow is very much the one we are creating right now.

We have seen that in quantum physics, the simple act of measurement at the subatomic level can cause a system to collapse from a quantum state of unlimited potential to a specific state of being either wave or particle. If you combine this finding with the concept of the butterfly effect, it is impossible to know what impacts a simple act of observation may have on future worlds. In fact, since paired particles can interact immediately across large distances, these simple acts of observation may have significant immediate effects that we can never understand. According to *Buddha Science*, any of these effects become part of our karma.

As we have stated, we know from biology the types of effect that organisms can have on their environment, but what effects do our social interactions have on our community and our society? Complexity studies are just beginning to reveal the larger societal patterns that we are subject to, and that we are creating today.

From the standpoint of the individual organism, we know the devastating effects that can occur when craving leads to addiction. But we also know that we can affect our karma throughout our lives by continued focus on the problems we face. Neuroplasticity studies have clearly shown that our brains are able to physically change at any age. The continued practice of meditation can create positive changes in mental focus and clarity, to help us more effectively deal with life's daily challenges. We build our karma, for better or worse, by every action we take.

BUDDHA SCIENCE

The third major observation of *Buddha Science* tells us that everything is impermanent. Everything is in the process of becoming, and this process will never end. Science tells us that chaos surrounds us, churning and unpredictable. It is in everything we see and know. Even the dimensions in which an object exists can change as a flat piece of paper becomes a paper ball.

The movement of time itself is relative, based on our movement through space and the gravitational forces we experience. Energy becomes matter; matter becomes energy, and the very particles that are the basic building blocks of physical existence wink in and out in unpredictable patterns. Or is it our concept of them that causes these particles to exist in the first place?

The cellular structure of your body continually changes, creating new cells while others die. And the genes themselves that perpetuate through many generations also eventually change and die. Communities change, becoming something new again every moment. Nothing is fixed and permanent. The thoughts you think are never the same, and the physical brain that processes your impressions is never the same from moment to moment. Only process exists.

Yet the very impressions we receive are constrained by our narrow perspective and the limitations of our recording apparatus. As *Buddha Science* points out, the world we think we experience is an illusion. Our bodies' sense organs record only a fraction of the physical world around us, and the most sophisticated devices we can create have only been able to measure 4% of the phenomena that science tells us must exist. Science can tell us very little about the remaining 96% of the universe that is known as dark matter and dark energy. States exist where things can be in two places at once, and more than one thing can occupy the same space. Cats and Buddhas can be both alive and dead, or neither.

The fourth observation of *Buddha Science* tells us that our impressions of the seemingly solid and real world around us are illusory. Our impressions of Reality are, at best, highly limited, and science has shown us time and time again that what we thought we knew was actually an illusion. Our senses only record a fraction of the world, and then our brain processes those limited sensations to present our conscious mind with a selected, preprocessed sample of those sensations, attaching memories, images, and emotions to it along the way. Yet we cling to our concept that the objects of our perception are solid and permanent, and our belief that matter and time and self are Reality. We continue to look at the map and believe it is the territory.

Science shows us that chaos and order are intertwined; yet we cling to order and push chaos away. Science tells us that space, time, matter, and energy are meaningless except relative to one another. Gravity does not exist as attraction between two bodies, but as a warping of the fabric of space and time itself.

Waves can be particles and particles can be waves, but when we measure them they seem to become some particular thing. Which state of being—the quantum state or the "normal" state—is real and which is illusion? Or are both some aspect of a larger Reality?

Biology tells us that individuals are merely a pattern within the larger process of life, but we still do not know exactly what life is or how it emerged. We have trouble even determining how to define intelligence when one-celled organisms with no brain or nervous system can exhibit seemingly intelligent behavior. We think we know what motivates us, and we think our choices are made consciously. But evolutionary psychology and neurobiology tells us that the very belief in conscious control is an illusion. Science tries to provide objective, falsifiable measurements of the things we study, but our understanding of the mind shows us that complete objectivity may never be possible. And when we come to studying the mind itself, how can we be objective if the mind is both the subject and object of our study?

Buddha Science tells us that our concepts of Reality are illusions, that although they may point the way toward Reality, a concept can never be Reality itself. Neither Buddhism nor science has the whole picture, but each may point to an aspect of the same Reality. There are other fingers that may point to the moon of Reality. We call them by such names as religion, metaphysics, and philosophy. Each has its own perspectives, and each may be right without negating the perspective of the others. Each of these views exists within Reality, and Reality cannot be constrained by the concepts we have of it, even if those concepts are illusions. Our error comes in believing that our concepts are Reality itself.

In his book about the illusion we call reality, Ralph Strauch presents a thought experiment about beings who perceive things exactly as we do, with the single exception that they are totally insensitive to temperature. Since they do not perceive temperature, they cannot create devices to measure it. To these beings, objects change size mysteriously, water becomes solid for no reason, and different things like snow and rain fall from the sky for no apparent reason.[232] Perhaps we are more like these beings than we want to admit. Perhaps dark matter and dark energy - which we believe make up 96% of the universe - are to us like temperature is to Strauch's fictional beings. Would our view of things change dramatically if we had some sense organs that detect dark energy? Is it possible that what science calls dark energy is the same thing that psychics and mediums and people with ESP are detecting through their senses? If we keep our minds open to such possibilities, we may see much more than we would if we shut down our options by clinging to more comfortable concepts. On the other hand, it is possible that we simply will never be able to see the whole elephant.

POOLING OUR THOUGHTS

So where do we go from here? If our goal is to understand what Reality is, as we stated at the beginning of this book, it may seem like a fool's errand. If concepts are not Reality, then how can it be possible to understand Reality through trying to form a concept of what it is? We can't. But what we can do is to use concepts as a starting point for understanding and then to get out of the way, embrace the stillness that surrounds us, and listen. Paradoxically, although concepts cannot take us face-to-face with Reality, they can help prepare us for that encounter, as long as we don't cling to these concepts when we encounter Reality itself. Once we cross the river, we need to leave our conceptual boat on the shore.

So with that in mind, I introduce a final concept, a metaphor that may be useful on your own path of discovery. I hope that his metaphor, along with the concepts and observations in this book, have helped you at whatever waypoint you may be along the path.

Let's begin by imagining that you are out for a walk in the wilderness and unexpectedly encounter a perfectly still pool of liquid. The pool is roughly circular but with many little ins and outs along the shoreline. The liquid in the pool is cool and dark, and there is no breeze to cause even the tiniest ripple. The surface of the liquid is an almost perfect, undistorted mirror of the sky above, yet there is some subtle vibrating pattern that seems to reflect massive hidden energy. Unsettled by such calm, vibrating power, you pick up a brown rock and chuck it into the center of the pool. The instant the brown rock touches the liquid, the pool vibrates more and a blue rock and a red rock immediately appear at other points in the pool, fly up out of the pool and land back on the shore. At the point, you realize there is something very odd about the pool.

Three perfectly circular waves spread out—one from the point where the brown rock entered the pool, and one from each of the points where the red and blue rocks appeared. Where these waves meet they form complex interference patterns. Because of the irregularities along the shoreline, when the three waves touch the shore there are many small circles that reflect back from various places around the shoreline. Each of these travels back across the pool and spawn yet more and more circles. And every time the circles cross each other, you see patterns of interference where the places they touch create mounds or troughs of liquid, which further complicate the patterns. You watch as the patterns become more complex and subtle, until finally everything settles into the faint vibrating pattern once more.

Fascinated by the process, you try it again with the red rock that emerged. This time, what looks like the same red rock immediately flies up from somewhere else in the pool, and the wave pattern that forms is equally complicated but completely different than the first one. Every time you try it, the results are similar but distinctive. Sometimes the same rock seems to emerge, sometimes something else appears. The wave patterns are distinctively different each time, but somehow very similar. Although these patterns slowly diminish, the original rock has never re-emerged. The actions you perform change the pool and the shoreline in ways that are sometimes subtle and sometimes more dramatic, but never predictably.

You look down into the pool to see what is happening, but the liquid is so dark and opaque that you can't see anything. Wary, you decide not to touch it. This is a good decision. If you could see into it, what you would see is that the pool is composed of pure energy, and time does not exist within it. The energy is everywhere at once. Anything that touches the pool (including your finger, if you had touched it) is immediately converted into pure energy. And the pool will only hold a certain amount of energy, so the moment an object is absorbed it must create and eject an object to keep the energy at the same level. Sometimes the objects it creates are identical to the ones it absorbs and sometimes they are completely different, as long as the energy levels match each other. The energy waves formed by object absorption and creation travel throughout the pool instantaneously, but they are slowed down on the surface due to the drag created by the world of objects. At the same time, the pattern that you see on the surface is not only due to the surface ripples, but also by the energy waves from inside the pool. But since the pool propagates energy instantaneously, the effects from within the pool are also happening instantaneously at the surface. That's why the pool vibrates, and why the red and blue rocks appear to emerge at the same time the brown rock entered the pool. The darkness of the pool makes it impossible to observe or describe it. This is a realm of undefined potential, always moving but always the same.

So now let's emerge from the fantasyland of the pool, and re-enter the fantasyland of our own existence. The pool is strangely familiar, but we may not be quite sure why. Well, for one thing of course, I designed the pool so it would be strangely familiar. But it's fair to ask what the heck I had to smoke before I came up with this crazy idea. Or at least, you might ask what it is I'm trying to say. All right, here goes.

The pool has much in common with what we have looked at in this book.

The surface can represent the conventional reality of our everyday lives—a reflection of our surroundings that is only a part of the whole pool. The whole pool is non-dual Reality, or if you prefer, the Tao, Brahman, God, the Void, or the quantum state. If you are of a scientific bent, you can call it dark energy. It includes not only what we can see on the surface, but the flowing, undefined potential of the energy below the surface. The patterns that form on the surface take shape through sensitive dependence on initial conditions, as we discussed in the chapter on chaos and complexity. In the words of *Buddha Science*, the patterns are created through causes and conditions, cause and effect, or karma. The patterns clearly show that everything is interconnected, both on the surface of the pool and within the flowing energy of the pool itself. We don't need to know exactly how it works to see that interconnection is occurring.

Let's apply the concept of the pool to the findings of quantum physics. First, we ask the question of whether the subatomic "particles" we think we see are in fact simply tiny little segments of waves that are so small they are at the limit of our ability to measure them. If so, it would not be surprising that they could interfere, add and subtract, and occupy the same space with other "particles." In short, it is easy to conceive that the "particles" we measure on the surface of this pool could experience superposition.

But what about simultaneity, what Einstein called spooky action at a distance? How can an object have an immediate effect on another object across the universe? In the case of our fantasy pool, the fact that the pool is made of pure energy serves this purpose. As we described in the chapter on physical science, if "space-time speed" is everywhere the same, anything that moves faster than the speed of light can be considered timeless. Everything happens at once, so the concept of time in such a system is meaningless. In the world of the pool, the red and blue rocks appear at the same time that the brown rock hits the water, while in our world, the paired subatomic particles communicate through this timeless energy medium.

What is this medium? What if the energy in the magic pool is equivalent to dark matter and dark energy, which comprises 96% of the known universe? Perhaps dark energy could be thought of as the energy in the pool, and dark matter is material that is in the process of transition out of the material world, slowed down by its exposure to the "negative energy" of mass. In other words, in this metaphor, dark energy is the content of the pool itself and dark matter is the vibrating patterns we see on the surface when that energy interacts with the physical world above the pool. When we observe this conversion happening on a random local basis, we see "particles" winking into and out of existence and interacting instantaneously. When we observe it happening on a massive, intergalactic scale, it shows up in our telescopes as dark matter.

Perhaps, what we conceive of as black holes is the equivalent of chucking the brown rock into the pool. Perhaps every action we take, every star that goes supernova, and every flap of a butterfly's wing causes dark energy waves to ripple through the universe, each one causing a pattern, whether large or small, but each inextricably linked to the Whole. Perhaps the karma created by our actions in this very moment has an immediate effect throughout the universe. Perhaps, anything is everything, with no dividing lines and nothing but an interconnected, dynamic pattern in the energy that is non-dual Reality.

As promised, we now return to the ideas of Amit Goswami and Peter Russell. Since there are no dividing lines, there is no reason to think that this vast pool of potential couldn't include consciousness as well. Perhaps dark energy is consciousness itself. Perhaps the very thoughts we have, and the awareness we develop, occur through constant interaction with this pool. Perhaps psychics and mediums can somehow tap into this pool and see things in a dimension that the rest of us do not detect. Or perhaps we all detect it, but we push it away because it seems crazy to us. Perhaps when our bodies are born our awareness rises from this whole, and when our bodies die, the awareness we think of as "self" will once again flow into this pool to re-emerge as a tree, an alien life, or as some far-away galaxy. Perhaps consciousness is the only true Reality, and everything else is simply a projection that is created by this imperfect instrument we call mind. Perhaps the mind of man is only the best instrument developed so far, and as we continue to evolve, the film in the projector will fall away and we, as a species, will begin to see a glimmer of the true undifferentiated light of consciousness.

But this is still only the way that we can envision it, because we are limited to concepts in three dimensions. We can develop equations, models, and maps that help us try to envision more dimensions, but when it comes down to it we are like the residents of Flatland. We simply cannot see what we cannot see. Perhaps if we could see the Reality around us from the outside, we would see a shape in dimensions that we can never conceive of from the inside, a shape that makes our ideas of time and space irrelevant. If the beings in this larger reality look like gods to us, then what about the beings in the next larger reality, and the next beyond that?

But of course, all this is only a concept.

As I might have mentioned once before, concepts are not Reality. At the beginning of this book, I spoke about the observation that we cannot measure such things as how you experience the color orange, your feelings when you see a sunrise, or how pecan ice cream tastes to you. We can talk about the studies of Reality and an intellectual understanding can enhance our appreciation of the world around us, but there are many aspects of Reality that can only be experienced. The momentary process that is experienced as "you" is imbedded in the endless flow of processes and phenomena, inextricably interconnected to everything that is. As you go about your daily life of cars and computers, cranes and calla lilies, let yourself experience a bit of the mysterious and miraculous web that surrounds you. Everyone comes to that experience in their own way, and I hope that by pointing out the incredible similarities that have been discovered through two very diverse approaches, it will aid you in your own experience. Echoing the words of American folk singer and troubadour Joe Crookston, I hope the words and concepts presented here will help you "turn yourself around again as part of the mystery."

Many brilliant scientists, observers, artists, novelists, and thinkers through the years have pointed their respective fingers toward the Reality they perceive, and the work will go on. Every book that has been written, every movie or recording that has ever been made, every conversation that has ever happened, all have contributed in some way to our collective striving to understand Reality. Or perhaps, all of these have created the reality we experience, a reality that is contained in and shapes the larger non-dual Reality that encompasses it all.

We all can be right. In fact, all of us are right because we are all part of that ultimate Reality. In this sense, it is impossible for anyone to be wrong. We are all experiencing the elephant from the inside. Collectively we are the elephant, so the reality we experience is of our own creation. We can choose to chop it up into little bits and fight for division in order to try and preserve our narrow belief in the rightness of our particular viewpoint, or we can open up to the complexity and richness of our collective vision.

Yes, embracing the whole is complicated, messy, and incomprehensible at times. Yes, in order to see the larger Reality we must see pain and injustice and destruction. We must see the dark side of ourselves in order to see the light. But we will never see Reality by refusing to see any part of it. That Reality includes science, metaphysics, religion, and Buddha *Science, of course.* It includes everything any conscious being has ever conceived or experienced, and much more that we may never know or understand. There is no end to the process, only the vibrating energy of our collective striving. But strive we must. As conscious beings imbedded within the Reality we seek to understand, it is our duty, and our privilege, to take this journey, to seek our place within that Reality. We are all connected through this journey of understanding.

In this book I have tried to capture a part of that collective striving toward understanding Reality. I hope that the concepts contained herein provide a reasonable map of two of those fingers—science and Buddhism—in their ongoing quest to point us toward the moon of Reality. I also hope that that this book has piqued your interested for further study and provides a starting point for that study.

Endnotes

1   David Bohm, *Wholeness and the Implicate Order* (New York: Routledge Classics, 2005), Kindle edition, location 141-44

2   Amit Goswani, *The Self-Aware Universe* (New York:Tarcher/Putnam, 1995), page 13

3   http://en.wikipedia.org/wiki/Blind_men_and_an_elephant, accessed 10/25/2014

4   Andrew Thomas, *Hidden in Plain Sight: The Simple Link Between Relativity and Quantum Mechanics*, ebook ISBN: 1469960796, 2012, location 125

5   B. Alan Wallace, *Hidden Dimensions: The Unification of Physics and Consciousness* (New York: Columbia University Press, 2007), Kindle edition, location 97-99

6   Dalai Lama, *The Universe in a Single Atom: The Convergence of Science and Spirituality* (New York: Morgan Road Books, 2005), page 269-270

7   B. Alan Wallace, *Hidden Dimensions*, page 42

8   Robert Kurzban, *Why Everyone (Else) is a Hypocrite: Evolution and the Modular Mind* (Princeton, New Jersey: Princeton University Press, 2010), page 14

9   Walpola Rahula, *What the Buddha Taught*, (New York: Grove Press, 1974), Kindle edition, location 265

10   Ibid., location 366

11   Steve Hagan, *Buddhism Plain & Simple* (New York: Broadway Books, 1945), page 2

12   Richard Hooper, *Jesus, Buddha, Krishna & Lao Tzu: The Parallel Sayings*, (Charlottesville, VA: Sanctuary Publications, 2007), page 5

13   Bhante Gunaratana, Jeanne Malmgren, *Journey to Mindfulness*, (Boston: Wisdom Publications, 2003)

14   http://secularbuddhism.org/, accessed 4/26/2014

15   Lawrence Krauss, *A Universe from Nothing: Why There Is Something Rather than Nothing* (New York: Atria Books, 2012), page 172

16   Amit Goswani, *The Self-Aware Universe*, page 16

17   David P. Barash, *Buddhist Biology: Ancient Eastern Wisdom Meets Modern Western Science* (New York: Oxford University Press, 2014), page 1

18   http://www.nytimes.com/2014/01/28/opinion/brooks-alone-yet-not-alone.html?action=click&contentCollection=Politics&module=MostEma iled&version=Full&region=Marginalia&src=me&pgtype=article&_r=0, accessed 1/30/2014

19   The Bible, Hebrews 11:1

20   David Bohm, *Wholeness and the Implicate Order* (New York: Routledge Classics, 2002), page 89

21   Norman L. Geisler, *Baker Encyclopedia of Christian Apologetics* (Ada, MI:Baker Books, 1999), page 446

22   Jerry Katz, editor, *One: Essential Writings on Nonduality*, (Boulder, CO: Sentient Publications, 2007), pages 41-45

23   Richard Hooper, *Jesus, Buddha, Krishna & Lao Tzu*, page 11

24   Marcelo Gleiser, *The Island of Knowledge*, (New York: Basic Books, 2014), page 16

25   Ibid, page 17

26   Dalai Lama, http://www.youtube.com/watch?v=bOpVrprggG0&index=2& list=PLOafJ4rP1PHwafTGL23zXK29knJsXMbMg, accessed 2/27/1014

27   Matthew Flickstein, *The Meditator's Atlas: A Roadmap of the Inner World*, (Somerville, Massachusetts: Wisdom Publications, 2007)

28   http://www.pnas.org/content/early/2013/08/08/1303312110.abstract, accessed 7/25/2014

29   Ralph Strauch, Joseph Chilton Pearce, Merna Strauch, Katarzyna Kozik, *The Reality Illusion: How You Make the World You Experience*, (Pacific Palisades, CA:Somatic Options, 2000), Page 39-40

30   Matthieu Ricard, Trinh Xuan Thuan, *The Quantum and the Lotus*: A Journey to the Frontiers Where Science and Buddhism Meet, (New York: Three Rivers Press, 2001), location 52

31   Dalai Lama, The Universe in a Single Atom: The Convergence of Science and Spirituality, (New York: Morgan Road Books, 2005), Kindle edition, locations 1623-26

32   Ibid, location 1629

33   Ibid, locations 76-78

34   Paul Lutus, http://arachnoid.com/is_math_a_science/, accessed 1/28/2014

35   http://www.exploratorium.edu/pi/history_of_pi/, accessed 1/31/2014

36   Yann Martel, Life of Pi, (Toronto: Knopf Canada) 2001

37   Gil Netter, Ang Lee, David Womark (Producers), Ang Lee (Director), Life of Pi, Twentieth Century Fox, 2012

38   http://www.piday.org/, accessed 1/28/14

39   http://www.numberworld.org/misc_runs/pi-10t/details.html, accessed 1/28/2014

40   http://education.jlab.org/qa/how-much-of-an-atom-is-empty-space.html, accessed 5/1/2014

41   Dan Hurley, *Breathing in vs. Spacing Out*, article in the New York Times Magazine, January 19, 2014, pages 14-15

42   Michael Speca, et al, *A Randomized, Wait-List Controlled Clinical Trial: The Effect of a Mindfulness Meditation-Based Stress Reduction Program on Mood and Symptoms of Stress in Cancer Outpatients*, in Psychosomatic Medicine, 2000, 62, pages 613-622

43 Matt Danzico, *Brains of Buddhist Monks Scanned in Meditation Study*, in BBC News, 4/24/2011

44 David Bohm, *Wholeness and the Implicate Order*, (New York: Routledge Classics, 2002) page 37

45 Ibid, page 51

46 Noson S. Yanofsky, *The Outer Limits of Reason: What Science, Mathematics, and Logic Cannot Tell Us*, (Cambridge, Massachusetts: The MIT Press, 2013) page 15

47 Ibid, page 8

48 Ibid, page 16

49 http://endless-satsang.com/advaita-nonduality-oneness.htm, accessed 4/21/2014

50 *Tao Te Ching* Chapter 14, translation to English by Gia-Fu Feng and Jane English

51 David Loy, *Nonduality: A Study in Comparative Philosophy*, (Amherst, New York: Humanity Books, 2012) location 149-152

52 Dalai Lama, *The Universe in a Single Atom*, locations 1854-58

53 Andrew Thomas, *Hidden in Plain Sight*, location 1489-90

54 Einstein did not use the name Flatworld, but spoke of flat beings on a spherical universe in the book *Relativity*; however, the origin of this concept seems to be Edwin A. Abbot, *Flatland: A Romance of Many Dimensions* (London: Seely & Company, 1884)

55 Albert Einstein, Robert W. Lawson, *Relativity: The Special and General Theory*, Amazon Direct Publishing, 2013, locations 1106-29

56 David Loy, *Nonduality*, locations 311-15

57 Dalai Lama, *The Universe in a Single Atom*, locations 760-63

58 http://www.thedhamma.com/whos_who.htm, accessed 10/26/2014

59 Stephen Bachelor, *Buddhism Without Beliefs* (New York, Riverhead Books, 1997), page 7

60   Steve Hagan, *Buddhism Plain & Simple*, page 7

61   Ibid, page 2

62   Jeffrey Grupp, speaking at UCLA 4-19-2006, ZenFlowerRadio.com, http://www.youtube.com/watch?v=tO-U5i3dc4k (accessed December 28, 2013)

63   http://www.goodreads.com/author/quotes/2167493.Gautama_Buddha, accessed 5/23/2014

64   Thupton Jinpa, http://www.youtube.com/watch?v=bXt8lbDZhVs&index=3&list=PLOafJ4rP1PHwafTGL23zXK29knJsXMbMg, accessed 2/28/2014

65   Steve Hagan, *Buddhism Plain & Simple*, page 9

66   Dalai Lama, *The Universe in a Single Atom*, location 1623-26

67   Swami Sivanada, Sure *Ways of Success in Life and God Realization* (Kuala Lumpur,Malaysia:Divine Life Society), Pages 94-99

68   Thich Nhat Hanh, *Being Peace*, (Berkeley, California: Parallax Press, 2005) page 51

69   Tong Shiu-sing & Hui Pak-ming, translation by Tong Shiu-sing & Janny Leung, *The Last Breath of Caesar*, from Physics World - http://www.hk-phy.org/articles/caesar/caesar_e.html

70   Archie Bunker was the main character in *All in the Family*, sitcom broadcast on CBS from 1971-1979

71   http://www.schiesshouse.com/beer_quotations_and_humor.htm, accessed 5/5/2014

72   http://en.wikipedia.org/wiki/Karma, accessed 2/5/2014

73   Stephen Bachelor, *Buddhism Without Beliefs*, page 37

74   Dalai Lama, *The Universe in a Single Atom*, location 1315-17

75   David Barash, *Buddhist Biology*, page 141

76   Steve Hagan, *Buddhism Plain & Simple*, page 58

77   John Briggs & F. David Peat, *Turbulent Mirror: An Illustrated Guide to*

Chaos Theory and the Science of Wholeness, (New York: Harper and Row, 1989) page 68

78   http://www.amazon.com/The-Outer-Limits-Colonel-Barham/dp/6302048931, accessed 2/6/2014

79   Walpola Rahula, *What the Buddha Taught*, location 718-21

80   Ibid, location 966-67

81   http://www.psychologytoday.com/basics/mindfulness, accessed 2/7/2014

82   Steve Hagan, *Buddhism Plain & Simple*, page 147

83   http://www.insightmeditationcenter.org/books-articles/articles/equanimity/, accessed 2/7/2014

84   Ibid

85   Karl Popper, *The Logic of Scientific Discovery* (Routledge Classics, Taylor & Frances e-library, 2005), page 66

86   Steve Hagan, *Why the World Doesn't Seem to Make Sense: An Inquiry into Science, Philosophy, and Perception,* (Boulder, Colorado: Sentient Publications, 2012), pages 47-48

87   Chava Frankfort-Nachmias and David Nachmias, Research Methods in the Social Science, (New York: Worth Publishers, 2007), pages 5-7

88   John Barrow, *The World Within the World,* (New York: Oxford University Press, 1988), pages 24-25

89   Jonah Lehrer, *The Truth Wears Off*, New Yorker Magazine, December 13, 2010

90   Amit Goswani, *The Self-Aware Universe*, page 96

91   Dalai Lama, *The Universe in a Single Atom*, locations 1410-12

92   Susan Blackmore, *Consciousness: A Very Short Introduction*, (New York: Oxford University Press, 2005), page 7

93   James Gleick, Chaos: Making a New Science, (New York: Open Road Integrated Media, 2011) page 41

94   Lisa Randall, *Higgs Discovery: The Power of Empty Space*, (New York: Ecco, HarperCollins, 2012) location 163

95   Lawrence Krauss, *A Universe From Nothing*, page 138

96   Andrew Thomas, *Hidden in Plain Sight*, page 132

97   Crutchfield etal., *Chaos*, Scientific American, December 1986

98   Jonah Lehrer, *The Truth Wears off*

99   Thomas S. Kuhn, *The Structure of Scientific Revolutions*, (London: The University of Chicago Press, 2012)

100 Noson S. Yanofsky, *The Outer Limits of Reason*, page 92

101 James Gleick, page 67

102 http://necsi.edu/guide/concepts/chaoscomplex.html, accessed 2/17/2014

103 http://www.wolframscience.com/reference/notes/971c, accessed 2/24/2014

104 http://www.stsci.edu/~lbradley/seminar/butterfly.html, accessed 2/24/2014

105 http://askville.amazon.com/phrase-Butterfly-Effect-originate/AnswerViewer.do?requestId=50157543, accessed 2/24/2014

106 http://necsi.edu/projects/baranger/cce.pdf, accessed 2/20/2014

107 James Gleick, *Chaos: Making a New Science* (New York: Open Road Integrated Media, 2011), page 168

108 Lawrence Krauss, *A Universe from Nothing,* page 153

109 James Gleick, *Chaos: Making a New Science*, page 311

110 www.sciencedaily.com/releases/2008/05/080507105644.htm, accessed 6/9/2014

111 James Gleick, *Chaos: Making a New Science*, page 221

112 Ibid, page 236

113 John Briggs & F. David Peat, *Turbulent Mirror*, page 95

114 Translators Gia-Fu Feng and Jane English, *Tao Te Ching*, Chapter 11 (New York: Vintage Books, 2011), page 13

115 http://www.necsi.edu/guide/study.html, accessed 2/27/2014

116 James Allen, opening remarks at *Fathoming Consciousness: Meaning and Measurement*, UM/ICAM Symposium, 2/14/2014

117 http://www.pbs.org/wgbh/nova/nature/emergence.html, accessed 5/8/2014

118 D. Bohm & B.J. Hilley, *The Undivided Universe: An Ontological Interpretation of Quantum Theory* (New York: Rutledge, 1993), page 8

119 John Briggs & F. David Peat, *Turbulent Mirror*, page 95

120 Sten Odenwald, www.astronomycafe.net

121 Fritjof Capra, *The Tao of Physics: An Exploration of the Parallels between Modern Physics and Eastern Mysticism*, (Boston, Massachusetts: Shambala Publications, 2010), page 172

122 Nagarjuna and Jay L. Garfield, The Fundamental Wisdom of the Middle Way: Nagarjuna's Mulamadhyamakakarika (New York: Oxford University Press, 1995), page 8

123 Ibid, pages 9, 134

124 Andrew Thomas, *Hidden in Plain Sight*, page 82

125 Albert Einstein, *Relativity: The Special and General Theory* (New York: Three Rivers Press, 1961), page 170

126 Fritjof Capra, *The Tao of Physics*, page 77

127 https://www.youtube.com/watch?v=xFqgSf7GDjI&list=UU0EO34a6lDzq rocgklRy_Fg&index=22, accessed 8/25/2014

128 Stephen Batchelor, *Buddhism Without Beliefs*, page 80

129 Bruce Rosenblum and Fred Kuttner, Quantum Enigma: Physics Encounters Consciousness (New York: Oxford University Press, 2011), page 76

130 Ibid, page 81

131 Nolson S. Yanofsky, The Outer Limits of Reason: What Science,

Mathematics, and Logic Cannot Tell Us (Cambridge, MA: The MIT Press, 2013), page212

132 http://www.sparknotes.com/biography/einstein/section9.rhtml, accessed 3/14/2014

133 https://medium.com/the-physics-arxiv-blog/d5d3dc850933, accessed 3/14/2014

134 http://www.forbes.com/sites/alexknapp/2012/09/06/physicists-quantum-teleport-photons-over-88-miles/, accessed 3/14/2014

135 http://www.cttbusa.org/preface/preface1.asp, accessed 3/14/2014

136 Andrew Thomas, *Hidden in Plain Sight*, page 7-8

137 John Briggs & F. David Peat, *Turbulent Mirror*, page 108

138 http://www.simonsfoundation.org/quanta/20140416-times-arrow-traced-to-quantum-source, accessed 8/19/2014

139 http://worldsciencefestival.com/events/higgs_boson_announcement/main, accessed 3/19/14

140 http://bicepkeck.org/, accessed 3/19/2014

141 http://www2.cnrs.fr/en/1525.htm, accessed 8/27/2014

142 David Loy, *Nonduality*, location 468-71

143 David Bohm, *Wholeness and the Implicate Order*, page 14

144 Ibid, page 37

145 Ibid, page 25

146 Translation by S. Mitchell, http://acc6.its.brooklyn.cuny.edu/~phalsall/texts/taote-v3.html, accessed 3/18/14

147 B. Alan Wallace, *Hidden Dimensions*, page 4

148 Peter Ward, *Life as We Do Not Know It* (New York: Viking Penguin, 2005), page 23

149 Ibid, page 12

150 http://www.oxforddictionaries.com/us/definition/american_english/life, accessed 3/21/2014

151 https://www.google.com/#q=define+life, accessed 3/21/2014

152 http://www.accesstoinsight.org/lib/authors/nyanaponika/wheel105.html, accessed 3/21/2014

153 David Barash, *Buddhist Biology*, page 43

154 John Briggs & F. David Peat, *Turbulent Mirror*, page 165

155 Norbert Wiener, *The Human Use of Human Beings: Cybernetics and Society* (Boston: Houghton Mifflin, 1950)

156 Nagarjuna and Jay L. Garfield, The Fundamental Wisdom of the Middle Way, page 106

157 Neil Shubin, *The Universe Within: Discovering the Common History of Rocks, Planets, and People* (New York: Pantheon Books, 2013), location 600-603

158 John Briggs & F. David Peat, *Turbulent Mirror*, page 68

159 David Barash, *Buddhist Biology*, page 57

160 Lawrence Krauss, *A universe From Nothing*, page 17

161 Rebecca Skloot, *The Immortal Life of Henrietta Lacks* (New York: Random House, 2010), page 217

162 Steve Hagan, *Buddhism Plain & Simple*, page 20

163 http://www.jameslovelock.org/page19.html, accessed 3/26/2014

164 Walpola Rahula, What the Buddha Taught, location 883-85

165 David Barash, *Buddhist Biology*, page 74

166 Ibid, page 139

167 http://media.hhmi.org/fittest/birth_death_genes.html, accessed 3/26/2014

168 http://www.lakehouse.lk/mihintalava/buddhism07.htm, accessed 3/28/2014

169   http://www.livescience.com/13363-7-theories-origin-life.html, accessed 3/28/2014

170   https://www.simonsfoundation.org/quanta/20140122-a-new-physics-the-ory-of-life/, accessed 3/28/2014

171   http://radio.foxnews.com/2014/01/29/audio-jeremy-england-explains-his-theory-of-how-life-began/, accessed 5/8/2014

172   Neil Shubin, *The Universe Within*, location 189

173   David Barash, Buddhist Biology, page 138

174   Ibid, page 24

175   David Barash, *Buddhist Biology*, page 20

176   http://www.youtube.com/watch?v=-8SORM4dYG8, accessed 9/15/2014

177   http://stopthebeetle.info/, accessed 4/2/2014

178   http://www.mnn.com/earth-matters/animals/photos/10-animals-that-are-bad-for-the-environment/disrupting-natures-balance, accessed 4/2/2014

179   Thich Nhat Hanh, *The Heart of the Buddha's Teaching*, (New York: Broadway Books, 1998), page 105

180   http://otec.uoregon.edu/intelligence.htm, accessed 4/3/2014

181   http://www.youtube.com/watch?annotation_id=annotation_497207&feature=iv&list=UUzWQYUVCpZqtN93H8RR44Qw&src_vid=HKZ2LtfDrmg&v=eXeygGxu8-8, accessed 3/16/14

182   http://www.nature.com/news/how-brainless-slime-molds-redefine-intelligence-1.11811, accessed 4/3/2014

183   Richard Golden, ed., http://sciphilos.info/docs_pages/docs_Huxley_T_Hcss.html, site visited 1/15/2014

184   Robert Wright, https://www.coursera.org/course/psychbuddhism. Special note: The author took this entire online course, beginning on 3/20/2014, and the information described is from lecture 1, part 4. The URL provided describes the course, which may not be available at a future date.

185 Wolfram Schultz, Paul Apicella, and Thomas Ljungberg, *Response of Monkey Dopamine Neurons to Reward and Conditioned Stimuli during Successive Steps of Learning a Delayed Response Task*, in The Journal of Neuroscience, March 1993, 13(3): 900-913

186 http://www.dummies.com/how-to/content/religion-for-dummies-cheat-sheet.html, accessed 10/27/2014

187 Steve Hagan, *Buddhism Without Beliefs*, page 17

188 Aura Glaser Robert A.F. Thurman, *A Call to Compassion: Bringing Buddhist Practices of the Heart into the Soul of Psychology* (Berwick, Maine: Nicholas-Hudson, 2005), location 1043-44

189 Robert Kurzban, *Why Everyone (Else) is a Hypocrite*, page 39

190 Ibid, page 68

191 Ibid, page 6

192 Ibid, page 53

193 Aura Glaser Robert A.F. Thurman, *A Call to Compassion*

194 http://www.psychologytoday.com/blog/the-second-noble-truth/201110/how-recognizing-your-death-drive-may-save-you, accessed 4/8/2014

195 UNODC, *World Drug Report 2013* (United Nations publication, Sales No. E.13.XI.6)

196 http://www.psychologytoday.com/basics/addiction, accessed 4/8/2014

197 http://intellihub.com/rat-park-experiment-shows-cultural-roots-drug-addiction/, accessed 4/8/2014

198 http://www.rightdiagnosis.com/p/psychological_addiction/symptoms.htm, accessed 4/8/2014

199 Robert Wright, https://www.coursera.org/course/psychbuddhism

200 http://www.oxforddictionaries.com/us/definition/american_english/brain, accessed 4/14/2014

201 http://dictionary.reference.com/browse/mind, accessed 4/14/2014

202 http://cogprints.org/6453/1/How_to_define_consciousness.pdf, accessed 4/14/2014

203 B. Allen Wallace, *Hidden Dimensions*, page 12

204 The Bridgekeeper, *Monty Python and the Holy Grail*, Dir. Terry Gilliam and Terry Jones, 1975

205 Stanislas Dehaene, *Consciousness and the Brain: How the Brain Codes Our Thoughts* (New York, Viking Penguin, 2014), Amazon Kindle Edition, location 700-704

206 Ibid, location 389-91

207 Walpola Rahula, *What the Buddha Taught*, location 697-706

208 Thupten Jinpa, Mind and Life XXVI, Day 2 pm video @1:42:20; http://www.youtube.com/ watch?v=JzKKuPmk95g&list=PLOafJ4rP1PHwafTG L23zXK29knJsXMbMg&index=5

209 Sankar K. Pal, Paul P. Wang, Genetic Algorithms of Pattern Recognition, (Boca Raton, Florida: CRC Press, 1996)

210 Chris Adami, *Evolutionary Approach to Artificial Consciousness*, guest lecture presented at the University of Michigan, 2/14/2014

211 http://gizmodo.com/an-83-000-processor-supercomputer-only-matched-one-perc-1045026757, accessed 4/17/2014

212 http://www.cnet.com/news/fujitsu-supercomputer-simulates-1-second-of-brain-activity/, accessed 4/17/2014

213 David J. Linden, *The Accidental Mind: How Brain Evolution Has Given Us Love, Memory, Dreams, and God* (Cambridge, Massachusetts: Harvard University Press, 2012), Amazon Kindle edition, location 1661-62

214 Stanislas Dehaene, *Consciousness and the Brain*, location 2441-2445

215 Susan Blackmore, *Consciousness*, page 17

216 Stanislas Dehaene, *Consciousness and the Brain*, location 160-62

217 Susan Blackmore, *Consciousness*, page 87

218 Stanislas Dehaene, *Consciousness and the Brain*, location 566

219 David J. Linden, *The Accidental Mind*, location 1152-54

220 Ibid, location 2899

221 Stanislas Dehaene, *Consciousness and the Brain*, location 2572-78

222 Rodney Smith, *Awakening*, page 176

223 Stanislas Dehaene, *Consciousness and the Brain*, location 1998-2000

224 Ibid, location 2002-2008

225 Ibid, location 2437

226 https://www.ted.com/talks/dan_dennett_on_our_consciousness
#t-972391, accessed 4/28/2014

227 Stanislas Dehaene, *Consciousness and the Brain*, location 1547-49

228 Peter Russell, From Science to God: A Physicist's Journey into the Mystery
of Consciousness, (Novato, California, New World Library, 2002), page 32

229 Amit Goswami, *The Self-Aware Universe*, page 141

230 Thupten Jinpa, Mind and Life XXVI, Day 2 pm video @1:57:55; http://
www.youtube.com/

231 Marcelo Gleiser, *The Island of Knowledge: The Limits of Science and the
Search for Meaning*, (New York: Basic Books, 2014), page 282

232 Ralph Strauch, *The Reality Illusion: How you make the world you experience*, (Pacific Palisades, CA: Somatic Options, 2000), page 119